03-ECO-2T

AF729482

Evaluation of WET Testing as an Indicator of Aquatic Health in Effluent-Dominated Streams: A Pilot Study

by:

Jerry Diamond
James Stribling
Tetra Tech, Inc.

2007

The Water Environment Research Foundation, a not-for-profit organization, funds and manages water quality research for its subscribers through a diverse public-private partnership between municipal utilities, corporations, academia, industry, and the federal government. WERF subscribers include municipal and regional water and wastewater utilities, industrial corporations, environmental engineering firms, and others that share a commitment to cost-effective water quality solutions. WERF is dedicated to advancing science and technology addressing water quality issues as they impact water resources, the atmosphere, the lands, and quality of life.

For more information, contact:
Water Environment Research Foundation
635 Slaters Lane, Suite 300
Alexandria, VA 22314-1177
Tel: (703) 684-2470
Fax: (703) 299-0742
www.werf.org
werf@werf.org

This report was co-published by the following organizations. For nonsubscriber sales information, contact:

IWA Publishing
Alliance House, 12 Caxton Street
London SW1H 0QS, United Kingdom
Tel: +44 (0) 20 7654 5500
Fax: +44 (0) 20 7654 5555
www.iwapublishing.com
publications@iwap.co.uk

© Copyright 2007 by the Water Environment Research Foundation. All rights reserved. Permission to copy must be obtained from the Water Environment Research Foundation.
Library of Congress Catalog Card Number: 2006928996
IWAP ISBN: 978-1-84339-768-7

This report was prepared by the organization(s) named below as an account of work sponsored by the Water Environment Research Foundation (WERF). Neither WERF, members of WERF, the organization(s) named below, nor any person acting on their behalf: (a) makes any warranty, express or implied, with respect to the use of any information, apparatus, method, or process disclosed in this report or that such use may not infringe on privately owned rights; or (b) assumes any liabilities with respect to the use of, or for damages resulting from the use of, any information, apparatus, method, or process disclosed in this report.

Tetra Tech, Inc.

This document was reviewed by a panel of independent experts selected by WERF. Mention of trade names or commercial products does not constitute WERF, EPA, Pima County Wastewater Management, nor the Arid West Water Quality Research Project endorsement or recommendations for use. Similarly, omission of products or trade names indicates nothing concerning WERF's, EPA's, Pima County Wastewater Management's, nor the Arid West Water Quality Research Project's positions regarding product effectiveness or applicability.

This project was co-funded by Pima County Wastewater Management and the Arid West Water Quality Research Project under EPA Region 9, Assistance Agreement XP-99926701.

ACKNOWLEDGMENTS

WERF acknowledges the commitment and dedication of the principal investigators and project subcommittee team in developing and refining this research project and in making the findings of this research relevant to WERF subscribers and the wastewater industry in general. Chad Barbour, Jonathan Bearr, and Lee Yarbrough of Tetra Tech assisted in conducting toxicity tests and analyzing data. Collin Hill, Adam Rettig, Carolina Gallardo, Jessica Garrish, and Chad Barbour, also of Tetra Tech, Inc. assisted in field sampling and macroinvertebrate sample processing. Advent Inc., Chadwick and Assoc., and Environmental Sciences, Inc. provided additional toxicity testing and field sampling assistance. The researchers are indebted to the cooperative assistance of the staff of six anonymous treatment facilities for being willing and cooperative participants in this project.

Report Preparation

Principal Investigators:
Jerome Diamond, Ph.D.
James Stribling, Ph.D.
Tetra Tech, Inc.

Project Team:
Jonathan Bearr
Marcus Bowersox
Henry Latimer
Tetra Tech, Inc.

Project Subcommittee

Thomas W. La Point, Ph.D., Chair
University of North Texas

Robin L. Autenrieth, Ph.D., Research Council Liaison
Texas A&M University

Rodney W. Cruze
City of Riverside, CA

Michael A. Lewis, Ph.D.
U.S. Environmental Protection Agency

Scott E. Belanger, Ph.D.
The Procter & Gamble Company

Gary Chapman
Paladin Water Quality Consulting

Daniel Santantonio, Ph.D.
City of Las Cruces, NM

Karmen King
Colorado Mountain College NMRI

Arid West Water Quality Research Project (AWWQRP)/Pima County

Karen Ramage, Program Manager

Richard Meyerhoff, Ph.D., Consulting Manager
CDM, AWWQRP

Christy Terhune, P.E.
Water Environment Research Foundation

Water Environment Research Foundation Staff

Director of Research: Daniel M. Woltering, Ph.D.
Program Director: Christy Terhune, P.E.

ABSTRACT AND BENEFITS

Abstract:

This project evaluated the quality of data needed to determine relationships between chronic Whole Effluent Toxicity (WET) test results and instream biological condition. A data quality objectives approach was used, which included several proposed measurement quality objectives (MQOs) that specified desired precision, bias, and sensitivity of methods used. Six facilities (four eastern and two western U.S.) participated in this study, all having design effluent concentrations > 60% of the stream flow. In accordance with a Quality Assurance Project Plan most of the facilities completed four quarters of chronic *Ceriodaphnia dubia*, and *Pimephales promelas* (fathead minnow) WET tests, and three quarters of *Selenastrum capricornutum* (green algae) WET testing following the most recent U.S. EPA methods. Several other WET tests were conducted to address MQOs including splits, duplicates, and blind positive and negative controls. Macroinvertebrate, fish, and periphyton bioassessments were conducted at multiple locations up and downstream of each facility following the most recent U.S. EPA Office of Water bioassessment protocols.

Test acceptance criteria were met for most WET tests, however, this study demonstrated the need to incorporate other MQOs in a full study (such as minimum and maximum percent significant differences and performance on blind samples) to ensure accurate interpretation of effluent toxicity. More false positives, lower test endpoint (i.e., higher toxicity), and more "failed" (non-compliant) tests were observed using No Observed Effect Concentrations (NOEC) as compared to IC_{25}s (concentration causing ≥ 25% decrease in organism response compared to controls). Algae tests often yielded the most effluent toxicity in this study, however, this test was most susceptible to false positives and high inter-laboratory variability. WET test results exhibited few relationships with bioassessment results, and could not usually predict instream effects even when incorporating actual effluent dilution. Neither frequency of WET non-compliance nor magnitude of WET were clearly related to differences in biological condition up and downstream of a discharge for the most part. Periphyton assessments were most able to discriminate small changes downstream of the effluent, followed by macroinvertebrate and then fish. The sampling methods used were robust but a full study should collect more field replicates up and downstream of each discharge to increase detection power. Macroinvertebrate and periphyton assessments together appeared to be sufficient to address project objectives. Fish assessments were not useful in this project as well but might be if more effort (and cost per site) were expended.

Benefits:

- Provides several data quality criteria for WET that should be considered along with test acceptance criteria to improve accuracy and defensibility of test results.
- Demonstrates the importance of data quality factors and endpoint calculation methods in determining whether an effluent "passes" or "fails" its WET limit.
- Provides data quality criteria and quality control measures for bioassessments that should be incorporated in any study.
- Provides analysis methods for identifying detrimental changes in biotic condition downstream of a discharge and methods to relate WET and biological results.

Keywords: whole effluent toxicity, bioassessment, chronic toxicity, data quality objectives

TABLE OF CONTENTS

The appendices for this report are available by request.

LIST OF TABLES

LIST OF FIGURES

LIST OF ACRONYMS

BRF	Biological Research Facility (Tetra Tech)
C_aCO_3	Calcium Carbonate
C.I.	Confidence Interval
CV	Coefficient of Variation
DQO	Data Quality Objective
IBI	Index of Biotic Integrity
IC_{25}	Concentration of a contaminant that inhibits organism's normal response (e.g., survival growth) by 25%
LOEC	Lowest Observed Effect Concentration
MHRW	Moderately Hard Reconstituted Water
MRPP	Multiple Response Permutation Procedure
MQO	Measurement Quality Objective
NMDS	Non-metric Multi-dimensional Scaling
NOEC	No Observed Effect Concentration
NPDES	National Pollutant Discharge Elimination System
NPE	Nonphenyl ethyloxylate
PMSD	Percent Minimum Significant Difference
PSE	Percent Sorting Efficiency
PTD	Percent Taxonomic Disagreement
QA	Quality Assurance
QAPP	Quality Assurance Project Plan
QC	Quality Control
RMSE	Root Mean Squared Error
RPD	Relative Percent Difference
SOP	Standard Operating Procedure
TAC	Test Acceptability Criteria
TEC	Test Evaluation Criteria
TU_c	Chronic Toxicity Units
WERF	Water Environment Research Foundation
WET	Whole Effluent Toxicity
WWTP	Wastewater Treatment Plant
YTC	Yeast Tetramin-Cerophyll

EXECUTIVE SUMMARY

A previous Water Environment Research Project (00-ECO-5) developed a study design to relate whole effluent toxicity (WET) test and biological assessment results for wastewater dischargers. A major conclusion of this project was that data collected using the most current and rigorous protocols are necessary to better define relationships between WET and biological conditions in streams. This study was designed using a Data Quality Objectives (DQO) approach, which included the evaluation of proposed measurement quality objectives (MQO) that were identified based on technical input from many scientists from federal, state, and private organizations. During the study design effort, several technical issues arose, which required investigation in this pilot study. Among these issues were to determine if DQOs and MQOs were achievable and determining appropriate biological assessment methods for certain types of aquatic systems, particularly arid west streams.

The goal of the current study was to determine how a comprehensive study should be designed to adequately address relationships between WET and instream biological condition. The objectives of this study were: 1) to evaluate the ability of biological assessments and WET tests to meet suggested DQOs and MQOs, 2) to determine the appropriate biological assemblages and evaluate the sampling protocols, and 3) to determine the minimum amount of WET and biological assessment data needed to adequately address project DQOs.

Six facilities (four eastern and two western U.S.) participated in this study, all having design effluent concentrations > 60% of the stream flow. In accordance with a Quality Assurance Project Plan most of the facilities completed four quarters of chronic *Ceriodaphnia dubia*, and *Pimephales promelas* (fathead minnow) WET tests, and three quarters of *Selenastrum capricornutum* (green algae) WET testing following the most recent U.S. EPA methods. Macroinvertebrate, fish, and periphyton bioassessments were conducted at multiple locations up and downstream of each facility following the most recent U.S. EPA Office of Water bioassessment protocols.

The six facilities examined in this study were located in five different states (four different EPA regions) and ranged in size from 1.5-185 million gallons per day (MGD) flow. All of the facilities had a design low flow effluent concentration of at least 62%. One of the facilities conducted WET tests monthly while the other five conducted tests quarterly.

Several MQOs for WET tests used in this study are already specified as test acceptability criteria. Where such criteria are lacking in the methods, criteria suggested by researchers or EPA were used. The latter include sensitivity MQOs such as percent minimum significant difference (PMSD) thresholds and precision MQOs such as coefficient of variation (CV) criteria for monthly reference toxicant test results. Bias MQOs were addressed through results of blind sample WET tests (positive and negative controls) and split tests. These included a blind method blank sample (false positive testing) and a blind toxic sample containing copper sulfate (false negative testing). In addition, MQO evaluation included split testing, in which the same effluent sample was tested simultaneously by two different laboratories, and duplicate tests, conducted in the same laboratory. Also, precision of reference toxicant test results over the course of a year were analyzed for each laboratory.

Field-based biological assessments are a series of methodological steps combined into a protocol. Therefore, MQOs are most appropriately established for each step. When individual performance characteristics are not applicable to some aspect of the assessment process, quality control procedures are mandated instead. Quality Control for bioassessments included standardized field and sample processing, standard operating procedures, as well as on-site audits and training. For the biological assessments used in this program, the steps were: 1) Field Sampling; 2) Laboratory Sorting and Subsampling; 3) Taxonomy; 4) Enumeration; 5) Data Entry; 6) Metric Calculation (i.e., Data Reduction; 7) Final Index and Site Assessment.

Two macroinvertebrate sampling methods, multi-habitat dipnet and Hester-Dendy multi-samplers, were examined as per the WERF study design for two facilities to determine the relative performance of each method and the need to use one or both in future studies. Macroinvertebrate sampling methods followed those in U.S. EPA's most recent Rapid Bioassessment Protocols. Fish sampling was conducted at least one day after macroinvertebrate sampling using backpack electroshockers and a team of four to five individuals. Larger streams required two teams for fish sampling. An average of eight man-hours was spent sampling each reach. Fish were typically identified in the field, by trained fish biologists. Voucher specimens were taken back to the laboratory to verify field identifications in a few cases. Periphyton sampling was completed by sampling habitats commonly used by periphyton (i.e., rocks, wood, plants) throughout the reach and scraping or brushing the periphyton from these items into a white plastic pan (Barbour et al., 1999). The composited sample from each reach was then poured into sample containers and preserved for transport to the appropriate laboratory. Periphyton data were standardized to a fixed volume prior to analysis.

Bioassessment data were analyzed using both multivariate and univariate statistical approaches. The former included similarity and non-metric multi-dimensional scaling analyses on assemblage taxa data. The latter used a multi-metric index analysis. Both analyses were used to assess precision of each bioassessment type as well as to test for differences in each assemblage between up and downstream samples.

WET Results

Test acceptability criteria were met in nearly all tests in this study. The few invalid tests were rerun in the same quarter. The project team observed a fairly wide range in documentation detail among laboratories in their WET test reports. As many of these reports were copies of what was sent to the regulatory authority, these differences may be important. Key issues noted were: 1) arithmetic errors in calculating *C. dubia* reproduction or fish weight, and number of broods per female in controls over the test, and 2) lack of information on sample holding times or sample temperature upon check-in. While most of the arithmetic errors were inconsequential in terms of the end results, it may be cause for concern if more widespread among WET testing labs.

Reference toxicant tests for each laboratory indicated compliance with WET protocols and endpoints that were within control limits. Three different reference toxicants were used by the six laboratories involved in this study. The frequency with which reference toxicant testing was conducted varied across laboratories from monthly to quarterly. One facility conducted only acute reference toxicant testing even though they conduct chronic WET tests monthly as part of their permit. The limited results here suggested that sodium chloride, the most common toxicant used in this study, may not yield desired endpoint precision among reference toxicant tests as compared to potassium chloride or nonphenyl ethyloxylate. There appeared to be more variability in *P. promelas* endpoints among laboratories that use sodium chloride than for *C. dubia*, which may indicate subtle differences in fish batch sensitivity to this chemical.

Effluent tests conducted by the six facilities yielded some chronic toxicity to at least one of the three test species. The *S. capricornutum* (currently *Pseudokirchneriella subcapitatga*) test yielded the most toxicity for five of the six facilities. No Observed Effect Concentration (NOEC) endpoints were often lower (i.e., indicated greater toxicity) than the corresponding IC_{25} endpoint (concentration above which at least a 25% decrease in control response was observed). Sublethal endpoints were often lower than endpoints based on survival in the case of *C. dubia* and *P. promelas* tests.

Most test controls exhibited satisfactory replication such that the recommended MQOs were met by most labs. This was particularly true for the *P. promelas* and *S. capricornutum* tests. The *C. dubia* tests displayed somewhat higher PMSDs with 76% of the *C. dubia* effluent tests reporting PMSDs < 40% (MQO was ≤ 37%). There was not necessarily a relationship between the labs that met the MQO for inter-test precision in reference toxicant tests and the MQOs for effluent test precision. For the *C. dubia* tests conducted in this project, an MQO of PMSD < 50% was very feasible (approximately 90% of all the tests conducted met this MQO).

Duplicate testing in each laboratory produced repeatable test results. However, nearly all samples conducted in duplicate testing by each facility happened to be non-toxic to both *C. dubia* and *P. promelas* (NOECs and IC_{25}s ≥ 100% effluent). This study can not address whether similar replication of results for these species tests is expected with effluents that have more inherent toxicity. The *S. capricornutum* test was duplicated in only two facilities; one with a toxic effluent and one with a non-toxic effluent, and both sets of tests exhibited fairly high agreement.

Out of the 12 chronic tests conducted in false positive testing, two *P. promelas* tests and all (i.e., two) *S. capricornutum* tests resulted in reported toxicity. However, the *P. promelas* false positives were based on NOEC values not IC_{25}s. The toxicity observed using the algae test and reconstituted moderately hard water in this study was one of several issues observed with results using this test method. *C. dubia* testing resulted in no false positives.

False negative testing (using a toxic blind sample) yielded some false negatives, particularly for the *C. dubia* test. Out of the 14 tests conducted in this analysis, there were a few that reported no toxicity in the sample using the *C. dubia* test in which the toxic concentration was designed to be at the high end of the test concentration series (75-90% of the sample). Examining both lethal and sublethal endpoints for each test, no false negatives were observed based on either the fish or the algae test, in which a toxic concentration was designed to be in the mid-range of the test concentration series. For the *C. dubia* test, half of the six labs reported the sample as non-toxic (NOEC or IC_{25} = 100%) and half the labs reported the sample as toxic (NOEC = 25-50% or IC_{25} 30-65%). If the *C. dubia* results observed in this project are representative of WET labs in general, these analyses suggest that data quality issues may be a significant factor affecting WET pass/fail interpretations for effluents that are toxic only at higher concentration.

Split testing of effluent samples indicated some disparate results in some cases. The MQO for this analysis was < 35 % difference in endpoints between laboratories for the split sample tests but this was not met in many of the tests, particularly for sublethal endpoints. The algae test resulted in the poorest agreement between the two labs and *C. dubia* had the best agreement. Generally, IC_{25} values between labs were within 50% of each other while NOEC values were often within 60% of each other.

The variability in endpoint values reported in split testing could translate to differences between passing and failing a WET test in some cases. Laboratory results for each facility were compared with the facility's WET limit or trigger concentration (based on design dilution instream) to determine whether a given laboratory result would be assessed as a "pass" (in compliance) or a "fail" (not in compliance). The algal growth test generated the most contradictory results between laboratories (Figure ES-1). For both types of statistical endpoints (IC_{25} or NOEC), four out of six split tests yielded contradictory interpretations between the laboratories. For the fish 7-day WET tests, the two laboratories reported results leading to contradictory interpretations half of the time, based on either the IC_{25} or the NOEC. The *C. dubia* 7-day WET test yielded the most reproducible results in split sample testing, with one out of six facility tests yielding contradictory interpretations using IC_{25} and two out of six yielding contradictory interpretations using the NOEC.

C. dubia and *P. promelas* survival endpoints resulted in predominantly passing results for most facilities when based on actual effluent dilution conditions (Figure ES-2). This was especially true when IC_{25} endpoints were used. Comparisons with limits or trigger concentrations based on design effluent dilution produced more "failed" tests than when compared to actual dilution concentrations as expected. Sublethal endpoints resulted in more "failed" tests than survival endpoints, using either IC_{25} or NOEC values, especially when compared to limits based on design effluent dilution.

Based on typical design flow permit limits, facilities had proportionally the highest "failed" test rate based on the *Selenastrum* test, followed by the *P. promelas* test, and the *C. dubia* test (Table ES-1). The "failed" test rate was higher using the NOEC as compared to the IC_{25} for all three test types. These results indicate that at least in some cases, the NOEC corresponded to less than a 25% decrease in organism response from controls.

Bioassessment Results

No one type of biological assemblage appeared to be most sensitive in this study, although periphyton exhibited effects at four of the six facilities. All three types of bioassessments (periphyton [four of six facilities], macroinvertebrates [two of six facilities], and fish [one of six facilities]) exhibited effects at different facilities (Table ES-1). Precision estimates, based on upstream sample metric scores, or multivariate similarity were highest for the macroinvertebrate dipnet bioassessment protocol, followed by periphyton and then fish. Lower fish bioassessment performance in this study indicates the need for more replicate samples at a given location and probably, the use of additional measures, such as fish condition or biomass to make this assemblage more sensitive. Given the relatively high cost and effort of this type of bioassessment compared to the others used, careful consideration should be given to using fish for studies with objectives similar to this pilot.

Both dipnet and Hester-Dendy gave reasonably similar results for the two facilities at which this comparison was made. The dipnet had somewhat better precision than the Hester-Dendy method, suggesting that the dipnet method may be preferred.

Laboratory processing of macroinvertebrate samples met MQOs for sorting efficiency (accuracy) and taxonomic accuracy (percent taxonomic disagreement) for five of the six facilities. The laboratory processing samples for one facility did not meet the MQOs initially but after corrective action, did produce satisfactory data quality.

Three of the six facilities exhibited significant decreases in pollution-sensitive metrics for periphyton while two facilities exhibited effects on macroinvertebrates and one facility had significant effects on fish (Table ES-1). Two facilities had apparent effects on both periphyton and macroinvertebrates and one facility exhibited effects on periphyton and fish. Based on these results, both periphyton and macroinvertebrate assessments together may capture most kinds of discharge effects.

Some of the most obvious changes in biological condition observed downstream of facilities in these analyses occurred at the one facility which did not have a viable upstream and used reference sites instead. While the reference sites used for this facility were believed to be appropriate according to facility staff and their consultant (the reference sites had been used previously by their facility for other projects), this result underscores the critical effect of using off-site locations as a surrogate for upstream conditions. Not all discharge situations will have readily available reference sites as in this project. For those situations, biological assessments at increasing distances downstream of the discharge may be a viable solution.

Most facilities had similar stream habitat up and downstream of the discharge as hoped but habitat quantity or quality did vary for one arid west facility (Facility #2), which may have confounded biological assessment results for that facility. A more comprehensive study would need even more careful selection of sites to ensure that habitat changes (not caused by the discharge) up and downstream are minimal or can be accounted for through other means. One such strategy is to compare abundance and distribution of biological indicators that are known to be sensitive to habitat changes observed, but not water quality changes that could originate from the discharge.

Relationships Between WET and Biological Condition

Few relationships were observed between the percentage of WET tests "failed" (i.e., not meeting WET limits) at a facility and the degree of biological change observed downstream in this study. This was true whether "failed" WET tests were based on actual or design flow conditions and was also the case when comparing the "fail" test rate for a given WET test species and its analogous bioassessment assemblage (e.g., *P. promelas* WET results and fish bioassessment). Since the "failed" test rate was as high as 58.3% for facilities in this study, these results were unexpected. However, given the results of the blind sample and split WET tests discussed previously, there is some uncertainty surrounding the quality of the WET data reported, and therefore the interpretation of WET test results.

No systematic bias was readily apparent with respect to either WET or bioassessment results. For some facilities, WET results would not predict instream effects yet significant decreases in the condition of certain biological assemblages was observed apparently unrelated to other factors. For other facilities, the opposite was true. MQOs tested and met in this project could, and perhaps should, be applied more generally to WET and bioassessment monitoring, particularly in a full study that examines relationships between WET and biological condition. As noted previously, all of the quality control testing in this study suggests that compliance with test acceptance criteria, while necessary and important, may not be sufficient for evaluating laboratory performance and data quality for these WET tests. Other quality control analyses (many of which are recommended by U.S. EPA and other regulatory agencies) are needed to ensure that correct test interpretations are made. For biological assessments, greater replication of biological samples and careful control of sampling locations to minimize confounding factors is recommended.

Table ES-1. Summary of Whole Effluent Toxicity and Bioassessment Results from the Six Facilities. + - No Detrimental Change Downstream, – = Detrimental Change Observed Downstream. C.d. = *Ceriodaphnia dubia*; P.p. = *Pimephales promelas*; s.c. = *Selenastrum capricornutum*.

	Percent Tests "Failed[1]					Bioassessment Results			
Facility	C.d.	P.p.	S.c.	Total	Total-Algae[2]	Periphyton	Macroinvertebrates	Fish	Total
1	25	25	100	40	20	+	+	+	0
2	25	50	100	44	37.5	–	–	+	2
3	25	75	100	58.3	44.4	–	+	+	1
4	0	25	100	36.3	12.5	–	–	+	2
5	0	0	100	33.3	0	+	+	+	0
6	0	20	0	6.7	10.0	–	+	–	2

[1] "Failed" tests are those WET tests that reported endpoints not meeting the facility WET limit or trigger concentration under either design or actual flow conditions.
[2]Total percentage of only *C. dubia* and *P. promelas* tests that "failed".

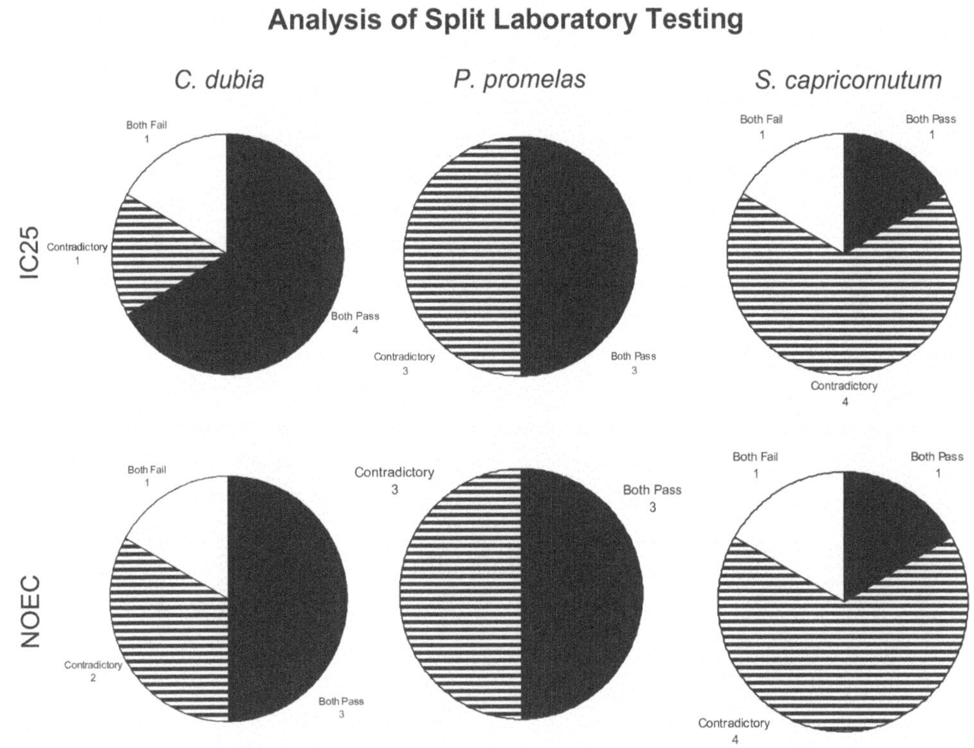

Figure ES-1. Comparison of Split Testing Results in Terms of Test Passing or Failing Permit Limits, With Respect to IC_{25} and NOEC Values Reported by the Laboratories for Each Type of Test: *C. dubia, P. promelas and S. capricornutum.*

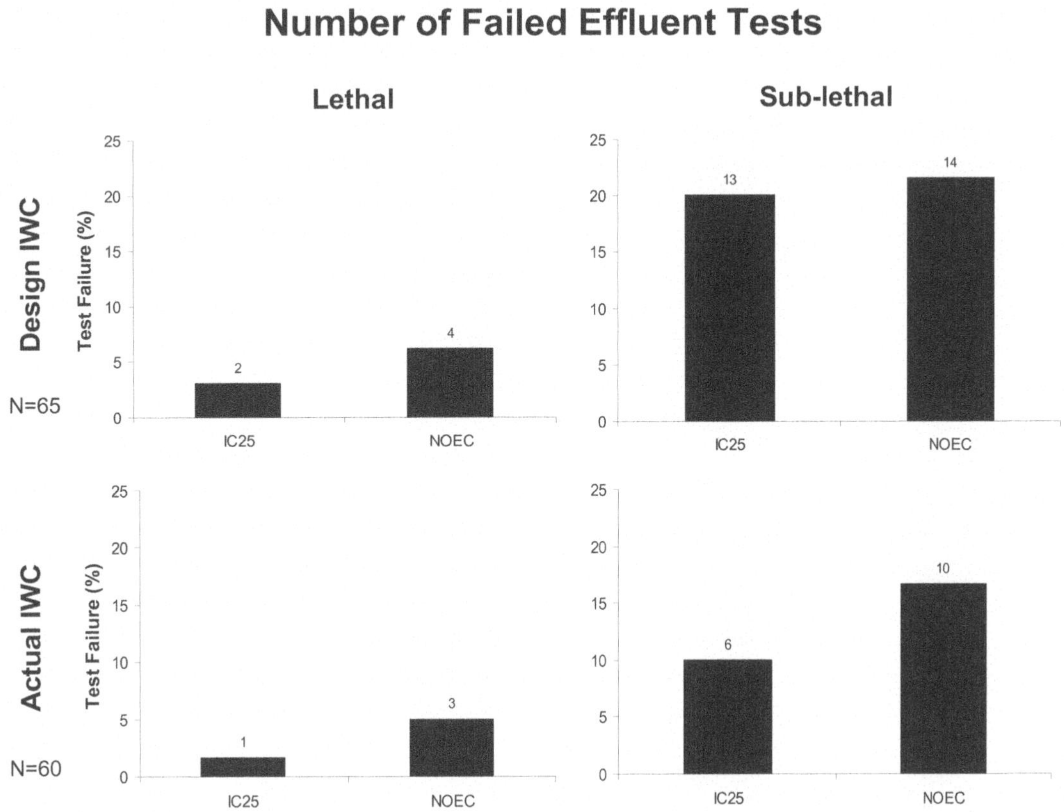

Figure ES-2. Comparison of Test "Failures" (i.e., Non-compliance With WET Limit) Based on Design and Actual Flow Conditions.

CHAPTER 1.0

INTRODUCTION

As part of the U.S. EPA's goal to ensure that designated beneficial uses of the nation's surface waters are met, whole effluent toxicity (WET) testing has been required in the National Pollutant Discharge Elimination System (NPDES) permits for many industrial and municipal dischargers. Given that WET tests are standardized (i.e., not site-specific) laboratory analyses that rely on responses of indicator species, the way in which WET test results are integrated in NPDES permitting and the ways that WET compliance limits are derived have been controversial issues. The crux of this controversy lies in the difficulty in relating WET test results to effects on biological communities in a receiving stream. Indeed, the problem of relating any laboratory-based result to actual field community responses has been one of the most intriguing yet challenging questions in ecology (e.g., Ankley, 1997; U.S. EPA 2000a) and applied biology (Waller et al., 1996; WERF projects 96-IRM-2, 97-IRM-3, 00-ECO-2, 03-ECO-1).

While there have been several previous attempts to examine relationships between WET results and biological condition [Eagleson et al., 1990; U.S. EPA 1991; Dickson et al. 1992, 1996], these studies used outdated test methods with unspecified data quality, and generally targeted sites known to have acute toxicity issues. Previous WERF research (WERF project 95-HHE-1; Diamond and Daley, 2000) examined this issue using WET and biological data compiled from over 90 sites across the U.S. that met strict quality control and other site selection criteria. Although some relationships between WET results and biological condition were observed in that study (predictably, when effluent dominated the stream flow under normal flow conditions and when acute toxicity was evident), results were somewhat inconclusive because: 1) few sites had all of the appropriate WET and bioassessment data needed for a rigorous statistical analysis; 2) WET test methods used in that study were still outdated in many cases; and 3) bioassessment methods and resulting data were not always of sufficient quality given more recent knowledge regarding the assessment of biological condition.

The current project is a continued effort based upon the study plan designed by this same research team in WERF project 00-ECO-5 (Diamond et al., 2002). In that study, the project team determined that several aspects of WET and biological assessment protocols lacked documented and accepted data quality characteristics with which reasonable data and measurement quality objectives (DQOs and MQOs, respectively) could be established (Diamond et al. 1996, Diamond et al. 2002). Based on that research, the project team recommended that a pilot study, using data collected by a few dischargers, should be implemented to validate the suggested DQOs and MQOs, prior to implementing a full study of the relationship between WET test results and instream biological conditions.

A major conclusion of the previous WERF project was that data collected using the most current and rigorous protocols are necessary to better define relationships between WET and biological condition in the receiving system. The current study was designed using DQOs and MQOs that were characterized using technical input from many scientists from federal, state, and private organizations. DQOs are qualitative and quantitative statements that clarify the intended use of the data, define the type of data needed to support the decision, identify the conditions under which the data should be collected, and specify tolerable limits on the probability of making a decision error due to uncertainty in the data (if applicable) (U.S. EPA, 2000a). These various criteria are developed by data users to specify the data quality needed to support specific decisions. Measurement quality objectives are quantitative criteria used to interpret the degree of acceptability or utility of the data in terms of meeting overall project objectives (U.S. EPA, 2000a).

During the characterization effort, several technical design issues arose, which required investigation in this pilot study. Among these issues were determining if DQOs and MQOs were achievable and determining appropriate biological assessment methods for certain types of aquatic systems, particularly arid west streams. The goal of the current study was to collect new data regarding method performance, for both WET and bioassessment methods, with the intention of addressing questions regarding data comparability and method selection for the definitive study.

This study was designed as a two-year project in which participating facilities would perform quarterly (at a minimum), U.S. EPA-approved short-term chronic WET tests (*Ceriodaphnia dubia*, *Pimephales promelas*, *Selenastrum capricornutum*) and instream bioassessments (macroinvertebrate, fish, algae) in the vicinity of their discharge point, as well as providing other prescribed data requirements. The results of this study are intended to provide answers to technical design questions raised by many researchers regarding ways to relate WET test results to instream biological condition. The results of this study should enable WERF and other interested parties to determine the most appropriate design for a definitive study.

1.1 Study Objectives

The objectives of this study were to evaluate the ability of biological assessments and WET tests to meet suggested DQOs and MQOs, to determine the appropriate biological assemblages and evaluate the sampling protocols, and determine the minimum amount of WET and biological assessment data needed to adequately address project DQOs.

When evaluating the quality of an environmental sampling program three important steps should serve as the focus of the evaluation. These steps are: 1) establishing DQOs and MQOs, 2) evaluating program design to determine whether it is sufficiently robust to meet study objectives, and 3) establishing quality control criteria or limits that can be used to document the appropriateness of the methods being used in the program. The quality of a particular dataset is some measure of the types and amount of error associated with the data. The data requirements for this project encompassed whole effluent toxicity testing and bioassessments. Therefore, it was imperative to collect sufficient QC data, as well as other types of data that could be used to identify actual performance of WET tests and bioassessment methods and compare that performance with desired MQOs.

1.1.1 Measurement Quality Objectives (MQOs) – WET

For WET tests, MQOs include quantitative statistics for test precision, bias, and sensitivity. Several MQOs for WET tests are specified as test acceptability criteria (Table 1-1). Where such criteria are lacking in the methods, criteria suggested by researchers or U.S. EPA were used. The latter include sensitivity MQOs such as percent minimum significant difference (PMSD) thresholds and precision MQO's such as coefficient of variation (CV) criteria for monthly reference toxicant test results. Bias MQOs were addressed through results of blind sample WET tests (positive and negative controls) and split tests. Results of WET tests conducted by laboratories in this pilot were analyzed in relation to the MQO's in Table 1-1.

1.1.2 Measurement Performance Criteria – Bioassessment

Field-based biological assessments are a series of methodological steps combined into an application as a protocol. Therefore, MQOs are most appropriately established for each step. When individual performance characteristics are not applicable to some aspect of the assessment process, it is indicated as such. For the biological assessments used in this program, the steps were:

1) Field Sampling
2) Laboratory Sorting and Subsampling
3) Taxonomy
4) Enumeration
5) Data Entry
6) Metric Calculation (i.e., Data Reduction)
7) Final Index and Site Assessment

Table 1-2 summarizes which MQOs were addressed in each step of a bioassessment protocol in this project. Certain types of MQOs, such as precision, could be addressed through quantitative criteria while other MQOs could only be addressed qualitatively, through quality control procedures.

1.1.3 Data Quality Objectives (DQOs)

Data quality objectives (DQOs) are qualitative and quantitative statements that clarify the intended use of the data, define the type of data needed to support the decision, identify the conditions under which the data should be collected, and specify tolerable limits on the probability of making a decision error due to uncertainty in the data (if applicable). DQOs are developed by data users to specify the data quality needed to support specific decisions. DQOs for WET testing and bioassessment includes measures for accuracy, power, and representativeness (Table 1-3). These and other information regarding performance measures for bioassessments are discussed in Section 2.0.

Table 1-1. Measurement Quality Objectives for Whole Effluent Toxicity (WET) Tests.

Test Acceptability Criteria	Toxicity Test Measurement Parameter	Measurement Quality Objectives
	Intra-test Parameters	
Yes	[1]Control survival (*C. dubia* and *P. promelas*)	≥ 80%
Yes	[1]Mean individual fish weight after a 7-day period	≥ 0.25 mg dry weight
Yes	[1]*C. dubia* offspring per adult female	≥ 15 within 7 days; ≥ 60 % controls have ≥ 3 broods within 7 days
Yes	[1]*S. Capricornutum* mean cell density in control	1×10^6 cells/ml
No	[2]Percent Minimum significant difference (PMSD) for *P. promelas* control growth (intra-laboratory)	≤ 35%
No	[2]PMSD for *C. dubia* chronic control reproduction	≤ 37%
No	[2]PMSD for *S. Capriconutum* control growth	≤ 20%
	Inter-test Variability	
No	[3]CV for monthly reference toxicant test endpoints	≤ 35%
No	[3]Split testing of effluent sample; CV for endpoints	≤ 35%
	False Positives/False Negatives	
No	[4]Split testing of method blank or positive control between labs: CV of endpoints	≤ 35%
No	[4]Associated reference toxicant test yields endpoint of the long-term average	± 35%

[1] Groethe, et al., 1996; U.S. EPA, 2000; U.S. EPA, 2002a.
[2] U.S. EPA, 2002b.
[3] Groethe, et al., 1996; U.S. EPA, 2000; Warren-Hicks, et al., 2000; Moore, et al., 2000.
[4] U.S. EPA, 2000b; Moore, et al., 2000.

Table 1-2. Summary of MQOs Used to Evaluate Biological Assessment Protocols in this Study.

Protocol Component	Assemblage Relevance: Benthos	Fish	Periphyton	Potential Error Source and Type	Proposed Error Control Procedure	Performance Characteristic Addressed	DQO or MQO
Field sampling	X	X	X	Patchiness of organism distribution (random error)	Large multihabitat sample	Representativeness	Standard protocol
	X			Degraded habitat/missing subhabitat (systematic error)	Consistent level of sampling effort/proportional distribution of sampling effort	Precision (intra-team and inter-team)	Metric value: ≤15% Metric score: ≤5% Index score: ≤15%
	X	X	X	Unique habitat or habitat type: nonrepresentative of site class or region (systematic error)	Stratification	Precision	Within-class variance ≤ among-class variance
	X	X	X	Recency of storm event/high flows (systematic and random error)	Postpone sampling for 2-3 days, duration of postponement depends on channel size	Representativeness	Standard sampling decision criteria
	X			Over- or under-sampling of subhabitats (systematic error)	Sampling effort (benthos) allocated to subhabitats in proportion to frequency of occurrence in defined sampling reach; sampling (fish) covers all habitat in sampling reach	Accuracy	Statement: Sample produces data representing biota that overall stream habitat has the capacity to support

Table 1-2. Summary of MQOs Used to Evaluate Biological Assessment Protocols in this Study (continued).

Protocol Component	Assemblage Relevance Benthos	Fish	Periphyton	Potential Error Source and Type	Proposed Error Control Procedure	Performance Characteristic Addressed	DQO or MQO
	X	X	X	Inconsistency of sampling (inter-team; random error)	Training, habitat definitions	Precision	Metric value: ≤15% Metric score: ≤5% Index score: ≤15%
	X	X	X	Inconsistency of sampling (intra-team, systematic error)	Training, habitat definitions	Precision	Metric value: ≤15% Metric score: ≤5% Index score: ≤15%
Sorting and Subsampling	X		X	Over- or under-representation of selected taxa (systematic error)	Fixed count/randomized subsampling, remove all organisms contained in aliquots	Accuracy, bias, representativeness	Standard subsampling protocol
	X			Non-representative taxa (systematic error)	Homogenization of sample, random selection of aliquots	Representativeness	Standard subsampling protocol
	X			Missed specimens (systematic error)	Sorter training/QC checks of sort residue	Accuracy (of sorter)	Sorting efficiency ≥90%
	X			Non-representative aliquot (systematic error)	Homogenization of sample; compare aliquot to total sample	Accuracy (of aliquot)	Relative percent difference (RPD) ≤5% (for composition metrics between aliquot and total sample)

Table 1-2. Summary of MQOs Used to Evaluate Biological Assessment Protocols in this Study (continued).

Protocol Component	Assemblage Relevance: Benthos	Fish	Periphyton	Potential Error Source and Type	Proposed Error Control Procedure	Performance Characteristic Addressed	DQO or MQO
Taxonomy	X	X	X	Mis-identification - taxonomic level (random error)	Designate more coarse (higher) taxonomic level as analytical requirement	Accuracy	Standard identification protocol
	X	X	X	Mis-identification - incorrect taxon (systematic error)	Use recognized taxonomy and technical literature	Accuracy	Standard verification protocol
	X	X	X		Use highly trained and experienced taxonomist; or recognized research expert		
	X	X	X		Identified and verified reference collection		
	X		X	Mis-identification - incorrect taxon (systematic error)	Re-identification of randomly-selected samples by 2nd laboratory (benthos), or duplicate samples collected at individual locations and split to different taxonomists	Precision	Percent taxonomic disagreement (PTD) ≤12

Table 1-3. Summary of Data Quality Objectives (DQOs) Use to Evaluate WET and Biological Assessment Protocols in this Study.

DQOs	Bioassessment	WET Testing
Accuracy	Frequency of metric agreement with assessment (based on chemical and physical data) of site impairment ≥85%.	Endpoint generated in performance evaluation (PE) testing ± 30% of mean value determined for sample. Determined through laboratory participation in PE testing.
Power	Detect ≥15% change in metric and index values 90% of the time with 90% confidence.	Ability to detect a 25% decrease in organism response 90% of the time with 90% confidence.
Representativeness	Sampling methods provide data that represent the actual stream community.	Results represent overall condition of effluent throughout the year.

CHAPTER 2.0

METHODS

2.1 Facility Selection Criteria

The project team surveyed over 40 wastewater dischargers with the objective of including in this study six facilities that met minimum criteria for meeting study objectives. Characteristics desired in selected facilities included: 1) effluent discharges to a wadeable freshwater receiving body (so that standardized bioassessment methods could be applied at the site); 2) no point or non-point source discharges within 1 km up and downstream of the discharge (to rule out other sources of stress as much as possible); 3) the effluent preferably makes up a substantial portion of the stream flow so that downstream effects, if observed, have a high likelihood of being associated with the effluent; 4) the facility was able to measure or estimate instantaneous flow conditions (to calculate actual dilution at time of WET testing), and 5) the facility is already required to conduct some WET testing, which could help defray test costs to WERF.

Although some facilities were able to support the project financially, inclusion in the project was not limited to those that were able to provide this support. Preference was given to those facilities that reported chronic toxicity within the 12 months prior to initiation of this investigation. Upon review of all submitted materials, and in concurrence with the Project Sub-Committee, the six facilities described below were chosen to participate in the investigation. Initially, three facilities located in the arid west and three located in the temperate east were selected. One of the arid west dischargers withdrew from the study in January 2005 and was replaced with an additional temperate east facility. Thus two arid west facilities (both having perennial upstream flow) and four temperate east facilities (three with perennial upstream flow and one with no upstream flow) participated in this study (Figure 2-1). Table 2-1 summarizes basic information for each participating facility relevant to this study. A brief description of each participating facility is included below.

2.1.1 Facility Descriptions

Facility 1. Facility 1 is one of the two municipal waste water treatment facilities located in the arid west, with a design flow of 36.3 million gallons per day (MGD) and a design dilution of 62.3% at 7Q10 stream flow. Facility 1 operates both primary (clarifiers) and secondary (trickling filters) treatment, with nitrifying trickling filters for ammonia removal. The effluent is disinfected by chlorination and residual chlorine is removed before discharge to the stream. The riparian land use surrounding the receiving stream is mixed, with light industry and open space dominating. The facility is required to perform quarterly *C. dubia* and *P. promelas* chronic toxicity tests. This facility was reportedly in compliance with its WET limit for the 12 month period prior to conducting this project. Bioassessments are not routinely performed at this facility. Figure A-1 (Appendix A, available by request) represents stream conditions as well as riparian habitat upstream of Facility 1, while Figure A-2 depicts representative conditions downstream of the discharge.

Facility 2. Facility 2, also located in the arid west, is a municipal waste water treatment facility that has a design flow of 185 MGD and contributes 98.6% of the downstream flow under 7Q10 conditions. Facility 2 is an activated sludge, advanced secondary treatment plant, with two separate treatment trains. One treatment includes nitrification/denitrification facilities and the other uses pure oxygen. Effluent is disinfected using chlorine and residual chlorine is removed before discharge to the receiving water. Riparian land use upstream of the facility is mixed use with light industry dominating, while downstream riparian land use consists of a mix of open space as well as light industry. Although the discharge from this facility is most of the downstream flow, upstream flow exists most of the time according to facility staff and could be sampled biologically in this study. However, flow volume was clearly much smaller upstream than downstream, even under average flow conditions. This facility currently performs monthly chronic *C. dubia* and *P. promelas* testing and was not out of compliance in the 12 months prior to beginning this project, according to facility staff. Bioassessments are routinely performed at this facility, including annual characterizations of the fish and macroinvertebrate communities within the receiving stream. Figure A-3 depicts the receiving water upstream of the facility, while Figure A-4 represents the receiving body downstream of the facilities discharge.

Facility 3. Facility 3 is a municipal activated-sludge waster water treatment plant located in the temperate eastern US, which, under design conditions, makes up 93% of the stream and has discharge of 2 MGD. Effluent is disinfected using chlorine and dechlorinated using sulfur dioxide gas before discharge to the receiving water. Riparian land use upstream of the facility is mixed with deciduous forest and open park space, while low-density housing is mixed with deciduous forest downstream of the facility. Under its current permit, this facility conducts chronic *C. dubia* toxicity tests quarterly. Facility 3 is not required to conduct fish or algal WET tests. This effluent was initially considered to be chronically toxic to *C. dubia* in the 12 months prior to conducting this project. It has since been determined that reported toxicity was due to laboratory error. Bioassessments are not routinely conducted in this receiving stream, but qualitative assessments of the fish community were conducted in 2004 and observed no significant difference in populations between up and downstream locations, relative to the discharge, according to facility staff. Figure A-5 is representative of the riparian and instream habitat downstream of Facility 3, while Figure A-6 depicts conditions upstream of the discharge.

Facility 4. Facility 4 is a municipal waste water treatment plant located in the temperate eastern U.S. that has a design flow of 1.5 MGD and a design effluent concentration dilution of 86.5% under 7Q10 stream flow. Facility 4 has secondary treatment, using rotating biological contactors for BOD and ammonia removal, after passing through several biological and oxidation processes. The effluent is chlorinated and then dechlorinated before discharge to the stream. Riparian land use up- and downstream of the facility is a mix of deciduous forest and low-density urban. This facility conducts chronic *C. dubia* and *P. promelas* testing upon permit renewal (every five years). The facility's compliance, with respect to toxicity in the 12 months prior to conducting this project, could not be determined due to the lack of toxicity testing in that period. However, the last set of tests (2001) indicated no chronic toxicity to either species. Bioassessments are not routinely conducted in this receiving stream. Figure A-7 depicts a sampling location just below the discharge and is typical of the receiving waterbody downstream of Facility 4. Figure A-8 is representative of the receiving waterbody upstream of Facility 4.

Facility 5. Facility 5 is a municipal waste water treatment plant located in temperate eastern U.S., with a design flow of 6.2 MGD that makes up 97% of the receiving stream under 7Q10 conditions. This facility has an activated sludge system coupled with a bio-filter for treatment of combined domestic and industrial wastewaters. The treatment process includes separation of solids, primary clarification, biological treatment, secondary clarification, effluent filtration, UV disinfection, and sludge handling prior to discharge to the receiving stream. The riparian land use up- and downstream of the facility is dominated by deciduous forest. The facility is required to conduct quarterly *C. dubia* testing only. The facility was out of compliance within the 12-month period prior to conducting this project. Bioassessments had not previously been conducted in the receiving stream. Figure A-9 depicts the sampling locations just below the discharge and is typical of the sites downstream of the facility. Figure A-10 is representative of the upstream sampling locations at this facility.

Facility 6. Facility 6 is an industrial facility located in temperate eastern U.S. that contributes 100% of the downstream flow to the receiving water body. Facility 6 discharges approximately 3 MGD of untreated contact and non-contact cooling water that does not get treated. The effluent is disinfected using chlorine and is dechlorinated before discharge to the stream. The upstream land use, which contributes minutely to the downstream flow, is a rural mix of agriculture and open space while the downstream riparian land use is a mix of agriculture and light industry. This facility is required to conduct quarterly *C. dubia* and *P. promelas* chronic toxicity testing. The facility was not out of compliance for the 12-month period prior to conducting this project (NOECs > 100% effluent). Bioassessments are not conducted routinely in the receiving stream but fish sampling was completed in 1999. Figure A-11 depicts one of the sampling locations on the receiving water, while Figure A-12 depicts one of the two reference sites sampled for this facility, due to the lack of upstream flow in the receiving stream.

2.2 WET Testing

As described in the study plan prepared under WERF project 00-ECO-5 (Diamond et al., 2002), WET tests were performed using *Ceriodaphnia dubia*, *Pimephales promelas*, and *Selenastrum capricornutum*. Sample collection, holding times, and WET test procedures followed the method requirements found in U.S. EPA's *Short-term Methods for Estimating the Chronic Toxicity of Effluents and Receiving Waters to Freshwater Organisms* (U.S. EPA, 2002a). A summary of test conditions for the species tested is shown in Tables 2-2 through 2-4. Most facilities supplied their own *Ceriodaphnia* and fathead minnow test data. Algal WET tests were conducted by two laboratories for all facilities. Reference toxicant tests in most cases were performed monthly for the species that a lab tested. However, certain laboratories conducted reference toxicant testing less frequently or conducted acute rather than chronic reference toxicant tests because of different NPDES permit requirements. Reference toxicant tests as well as effluent test data were evaluated in accordance with laboratory control chart data and reviewed by the project team to document acceptability in relation to U.S. EPA's test acceptability criteria (U.S. EPA 2002a) and project MQOs.

An important aspect of quality control used in WET testing pertains to the collection of physicochemical data of the test concentrations to ensure that no parameters fall outside of acceptable ranges set by U.S. EPA (U.S. EPA, 2002a). This information was recorded daily on test sheets by each laboratory and reviewed by the project team. All test data were entered into spreadsheets for various analyses described below. To ensure accuracy, all data were checked by QC staff using the original laboratory datasheets.

Another important aspect of quality control performed during this project was the inclusion of blind testing, duplicate testing, and split sample testing with another participating laboratory. Each of these analyses was conducted once during the project and used to address the level of agreement (relative percent of false positive/false negative results), and precision for each type of WET test.

Two different blind tests were conducted in this study. The first test used a method blank that consisted of reconstituted moderately hard water prepared at Tetra Tech's Biological Research Facility (BRF) using the protocol in U.S. EPA (2002a) and deionized water (Barnstead E-Pure, 18 mohm resistivity). A single batch of the blank sample was prepared and then split into six subsamples (one for each facility), placed in clean nalgene cubitainers and then shipped overnight on ice to the respective laboratories. All facilities, with the exception of Facility 3 (due to financial constraints), were able to conduct testing using the blank sample. The second blind sample was a positive control consisting of a copper sulfate solution (6.2 mg/L). Reagent grade copper sulfate was dissolved in deionized water at Tetra Tech's BRF, thoroughly mixed, and then portioned out into nalgene bottles which were then shipped to respective labs. All facilities were able to conduct testing using the positive control sample. To ensure accuracy of test setup, each facility was given instructions on how to dilute each blind sample and prepare test concentrations (Appendix B, available by request). False positive rate of WET test measurements was examined through the use of the sample blank results. False negative rate, bias, and precision of WET test measurements were evaluated by examining positive control results, as this sample had a known effect range for a given species and endpoint.

Duplicate testing was performed by conducting side-by-side testing of one set of WET tests for each facility. The objective of duplicate testing was to determine the intra-laboratory precision of each WET test type.

All facilities conducted split effluent tests once in this project. One test was conducted by the facility's laboratory and a split test on the same set of samples was conducted by another laboratory for that facility. Tetra Tech's BRF served as the split laboratory for all *C. dubia* and *P. promelas* testing, except Facility 4. ESI Inc., in Hampton, NH, conducted split *S. capricornutum* WET tests for Facility 6, because Facility 6 conducted their own algae tests. The objective of conducting split effluent sample testing was to estimate the interlaboratory precision of each WET test.

2.3 Bioassessments

2.3.1 Field Sampling

Bioassessments were conducted at each facility for fish, macroinvertebrate, and periphyton assemblages. Protocols for each type of bioassessment were derived from U.S. EPA's Rapid Bioassessment Protocols (Barbour et al., 1999) and included in a Quality Assurance Plan (Appendix C, available by request). The study plan called for stream sampling to be performed within three kilometers above the discharge and within three kilometers below the edge of the chronic mixing zone or the dilution point dictated by the facility's compliance limit for WET testing (See Figure 2-1). This was to ensure that downstream effects, if they occur, were properly identified with respect to either water quality or habitat effects of the discharge, or other physicochemical effects independent of the discharge. This design presupposed a viable upstream for each facility. To supplement biological data, habitat assessment data and instream physicochemical data (pH, dissolved oxygen, temperature, and conductivity) were collected simultaneously, also adhering to current Rapid Bioassessment Protocols (Barbour et al., 1999).

Biological samples were taken at multiple points throughout a sample reach for each assemblage (Figure 2-2). For macroinvertebrate and periphyton assessments, the sampling reach was defined as 30 times the wetted stream width, or a minimum of 150 m. For fish assessments, a stream reach was 12 times the wetted width. In this study, all facilities could be sampled upstream of the discharge with the exception of Facility 6. This facility discharges to an effluent dominated stream with no flow above the discharge point. At this facility, samples were taken from two reference streams within the same ecoregion that have been used previously by this facility. At Facility 5, 90-meter reaches were used rather than a larger reach as proposed, due to a dredging operation located downstream, which would have fallen within two sampling sites given the reach length specifications.

Data quality related to bioassessment procedures was addressed by consistent performance of procedures documented in the SOPs (Appendix C, available by request) and by the documentation of performance characteristics in this project. All individuals involved with field sampling received in-depth training by the project team prior to sampling. While in the field, a QC officer audited all sampling as well as all data recorded on field data sheets to ensure data completeness and accuracy.

To address data representativeness (i.e., data collected for a given site being truly representative of the full assemblage that the stream has the ability to support), macroinvertebrate and fish samples were collected using a multi-habitat technique to ensure that the samples gathered depicted the biota the existing habitat had the capacity to support (Barbour et al., 2006 [in review]). Periphyton were sampled from a single habitat in each reach that provided the most optimal conditions for periphyton growth.

Macroinvertebrates were sampled using a 0.3 meter 595 μm mesh D-frame net by kicking the substrate or jabbing the net into all available instream habitats (Barbour et al., 1999). In addition, artificial substrates (Hester-Dendy multi-plate samplers) were used to sample macroinvertebrates at two facilities (Facility 5 and 6). These samplers provided a total of $95 in^2$ ($613 cm^2$) of surface area that could be colonized at each location. Use of Hester Dendy samplers avoids the need to have similar habitat among sites being sampled, and theoretically allows interpretation of strictly water quality effects (Rosenberg and Resh, 1982). For all sites, with the exception of Facility 6, macroinvertebrate sampling occurred at three sites above and three sites below the discharge, typically separated by 15-25 meters (with the exception of sites described above). At Facility 5, three samplers were placed upstream of the discharge and three below. At Facility 6, two samplers were used for the "upstream" location; one for each of the two reference streams. The samplers were kept instream at each facility, for a period of six weeks prior to retrieval (Barbour et al., 1999).

Fish sampling was conducted at least one day after macroinvertebrate sampling using backpack electroshockers and a team of four to five individuals. Larger streams required two teams for fish sampling. An average of eight man-hours was spent sampling each reach. Fish were typically identified in the field, by trained fish biologists.

Periphyton sampling was completed by scraping or brushing the periphyton from rocks, wood, or plants into a white plastic pan (Barbour et al., 1999). A measurable volume of algae was sampled in each reach. The composited sample from each reach was then poured into sample containers and preserved for transport to the appropriate laboratory. Periphyton samples were sent to Dr. Rex Lowe at Bowling Green University (OH) for algae enumeration (cell density, chlorophyll *a*) and taxonomic identification for Facilities 1 through 5. Advent Environmental Services processed periphyton samples for Facility 6. Periphyton data were standardized to a fixed sample volume prior to analysis.

Fish were enumerated and identified on site for each sample. Voucher specimens were taken back to the laboratory to verify field identifications in a few cases.

2.3.2 Macroinvertebrate Laboratory Sample Processing

Macroinvertebrates were sorted and subsampled by either the project team (Laboratory A: Facilities 1-5) or by a separate laboratory (Laboratory B: Facility 6) using the SOP contained in the project Quality Assurance Project Plan (QAPP; Appendix C) and derived from protocols in Barbour et al. (1999). The subsampling procedure used a Caton gridded screen (Barbour et al. (1999) to allow discrete "aliquots" of the sample material to be randomly selected, and removed from the grid tray for picking. The target number of organisms for identification was 200 ± 20% (160-240). Samples with less than 160 organisms in the entire sample were noted as such.

Bias of invertebrate laboratory sorting and subsampling was evaluated by characterizing percent sorting efficiency; PSE is used to evaluate the effectiveness of a laboratory or of a single sorter in finding and removing specimens. After the initial sorting effort conducted by Lab B, the remaining sample fraction (pickate) was rechecked by Lab A (as external QC) for 10% of the samples. The QC laboratory performed sort re-checks under the same conditions as were used in Lab B, no magnification (naked eye only) in the case of the dip-net sample, and with magnification (<3x) in the case of the Hester Dendy sample. If organisms were found, they were removed and placed in a vial containing approximately 80% ethanol, and labeled with all of the original label data, and designated "pickate recoveries". The number of organisms recovered from the pickate during the secondary recheck (external QC) was used to calculate the secondary percent sorting efficiency (PSE_2), as follows (Hill et al., 2005):

$$PSE_2 = \left(\frac{A}{A + B_1}\right) * 100$$

A = number of organisms originally found

B_l = number of organisms found in external QC

The measurement quality objective (MQO) for sorting and subsampling bias was: <10% of the samples with <90% PSE.

Taxonomic precision was evaluated by calculating percent taxonomic disagreement (PTD) and percent difference in enumeration (PDE) of the identification results from approximately 10% of identified benthic samples. Identifications for Lab A were identified by an independent taxonomic laboratory (Lab C), and identifications for Lab B were identified by a taxonomist within their facility. Both laboratories use the most appropriate and up-to-date technical literature. Taxonomy was performed to specified hierarchical levels, mostly to genus, some to species and others to higher levels (i.e., tribe, family, subfamily, order, or class). Approximately 10% of the samples processed by Lab B were randomly chosen for re-identification. Once the primary identifications were completed for the two samples, the vials and slides containing specimens were shipped to an independent taxonomic lab (Lab C), for re-identification. Another aspect of sample processing that is related to taxonomic identifications is enumeration, or the direct counts of individuals in sample, both in total and separated by individual taxa.

Precision of taxonomic identification was assessed by comparing genus-level taxonomic results from two independent taxonomists. A side by side comparison between taxonomic results was performed for both of the re-identified samples. The process entailed examining both taxa lists for discrepancies between them, including straight disagreements (i.e., two different genera), hierarchical disagreements (i.e., family level identification vs. genus level), and numerical disagreements (i.e., different numbers of organisms listed under each taxon), and then determining the number of agreements between each list. Using the number of agreements, percent taxonomic disagreement, or PTD (Stribling et al., 2003) was calculated using the equation:

$$PTD = \left[1 - \left(\frac{Comp_{pos}}{N}\right)\right] \times 100$$

Where $Comp_{pos}$ is the number of agreements, and N is the total number of organisms in the larger of the two counts. The lower the PTD value, the more similar are sample taxonomic results, and the greater is the overall taxonomic precision. The MQO for precision of taxonomic identifications are: A) PDE <5% (i. e., there will be <5% difference in final organism counts between re-identification results for a given sample), and B) PTD $\leq$15% (i. e., rate of taxonomic error will be <15% for a given sample).

2.4 Sampling and Testing Summary

Table 2-5 summarizes all WET testing completed for each facility that participated in this research project. Five facilities were able to complete at least four quarters of testing with *C. dubia* and *P. promelas*. Facility 5 did not complete the first two quarters of *P. promelas* testing due to miscommunications between this Facility and its contract laboratory. This facility is not required to perform this test according to its permit. Most facilities were unable to complete the *S. capricornutum* test in the first quarter because the actual testing phase of the project started near the end of the first quarter for most facilities. None of the Facilities are required to conduct this test according to their permits. Laboratories conducting *C. dubia* tests for all size facilities currently culture their own *C. dubia*. Fathead minnows were either cultured (Lab for Facility 1) or purchased through various vendors (Labs for Facilities 2, 3, 4, 5, 6). All three types of bioassessments (periphyton, macroinvertebrates, and fish) were completed within the respective state index periods at all facilities.

2.5 Data Analysis

2.5.1 WET Testing

All WET test data were reviewed by the project team to determine whether each test met the acceptance criteria. The number and percentage of invalid tests (based on acceptance criteria and MQOs) was compiled for each test type and all labs combined. Effluent test precision was analyzed by computing percent minimum significant differences (PMSD, U.S. EPA 2002b) for the *C. dubia* and *P. promelas* tests in the study.

In addition, WET test data and quality control test data were subject to several other analyses to determine overall WET variability for each test type and compliance with stated DQOs and MQOs. WET analyses conducted were as follows:

Chronic WET test endpoints are commonly calculated using two different analysis approaches. In the first approach, organism responses in each effluent concentration are compared to control responses using a hypothesis-testing (e.g., Analysis of Variance [ANOVA]) approach, which yields a No Observed Effect Concentration (NOEC) and a Lowest Observed Effect Concentration (LOEC). These endpoints are derived by testing statistically the response in each effluent concentration as compared to controls (e.g., Dunnetts Test or Wilcoxon Rank Sum Test) and therefore, both the NOEC and LOEC correspond to test effluent concentrations used in the WET test.

A second approach uses a probit analysis or similar regression technique to identify the effluent concentration which, on average, will yield a specified difference from the control response. In WET testing, this specified difference is typically a 25% reduction in response as compared to controls (e.g., a 25% decrease in growth) and is referred to as an IC_{25} – the concentration of effluent that results in a 25% inhibition in response as compared to controls. Unlike an NOEC or LOEC, the IC_{25} approach yields an endpoint that does not often correspond to a specific concentration actually tested in WET testing, but uses the entire concentration-response relationship. It also typically allows the calculation of 95% confidence intervals around the IC_{25}, a statistic that an NOEC or LOEC does not readily provide. Currently, most NPDES WET limits are based on NOEC endpoints, however, there is increasing interest and use of IC_{25} endpoints, as well (Chapman et al., 1996). Given the prominence of both types of statistical endpoints in WET testing, this study examined MQOs and other aspects of WET data in light of both IC_{25} and NOEC approaches.

Reference Toxicant Test results of each lab and test type were compiled and analyzed to determine the long-term endpoint average and its variability. These statistics were computed using U.S. EPA's recent methodology to determine precision of reference toxicant test data for each test type (U.S. EPA 2002b).

Positive control results were tabulated for each laboratory and evaluated to determine the false negative error rate for each test type and each lab.

Negative control results were tabulated and evaluated to determine the false positive error rate for each test type and each lab.

Split sample testing results were evaluated to determine endpoint variability between labs for each test type and endpoint evaluated.

Duplicate test results were evaluated to determine endpoint variability within labs for each test type and endpoint measured.

Prediction of effluent effects instream: This set of analyses categorized each test as either a "pass" or a "fail". Test categorizations were defined based on both design and actual effluent dilution conditions for each test. In each case both No Observed Effect Concentration (NOEC) and IC_{25} (concentration at which there is a 25% decrease in organism response as compared to controls; U.S. EPA, 2002a) values for survival and sublethal endpoints were compared with design or actual effluent dilution to determine whether a test was ranked as a "pass" or "fail". Both the frequency and the magnitude of "failed" tests under design and actual dilution conditions were examined for each facility to determine the likelihood of biological effects instream.

2.5.2 Bioassessment Analyses

Several types of analyses of bioassessment data were performed to evaluate a) the degree to which MQOs were met; b) whether different types of biological assessments yield similar and therefore redundant results, and c) whether both macroinvertebrate assessment protocols used in this project are similar in terms of their resultant data quality and ability to meet desired DQOs and MQOs. (See Figure 2-3)

Multivariate Analyses. Analyses of bioassessment data took two general forms. One line of analysis examined similarity of species composition and biological metrics between samples or reaches using a multivariate approach. The degree of similarity observed among upstream samples for a given assemblage is indicative of the precision of the method and identifies the degree of difference needed to detect a significant change in the assemblage downstream. The degree of similarity between up and downstream samples for a given facility indicates the degree of biological change in an assemblage. Further analyses were conducted to determine what type of change was observed and whether much or all of the change could be attributed to changes in flow between up and downstream at a given site.

Similarity analyses have the advantage of being able to easily display the degree to which samples are similar in terms of their biological structure (composition and relative abundance of taxa) and can be used for any assemblage. It also has the advantage of being able to evaluate the similarity of several metrics combined among samples. This was desirable in this project because we could not rely on the state Index of Biotic Integrity (IBI or other index approach) since the same sampling and sample processing methods could not always be in used in this project. The researchers did, however, make an attempt to evaluate the same metrics the state uses in their bioassessment programs.

The percent abundance of fish, macroinvertebrate, or periphyton taxa in each sample was used in separate ordination analyses using non-metric multidimensional scaling (NMDS) to determine similarity among samples based on Bray-Curtis similarity of taxonomic composition for each assemblage (McCune and Grace, 2002). Multiple response permutation procedure (MRPP) (McCune and Grace, 2002), a statistical technique for testing differences between groups, was used to examine the heterogeneity within groups and differences among groups and to determine significance of results. For macroinvertebrates, several common metrics used by each state were calculated using the Ecological Data Application System (EDAS) database for each benthic sample. These metrics included the best candidate metrics as described in Barbour et al. (1999).

Similarity metric values were calculated using Euclidean Distance (McCune and Grace, 2002) and these similarities were compared both among upstream sites above each facility (to calculate precision) and between upstream and downstream for each facility using NMDS and MRPP. The two different macroinvertebrate sampling techniques, dip net and Hester Dendy, were also compared based on taxonomic composition. Blocked MRPP tests were used to test the differences between paired upstream-downstream samples for each method and between upstream samples alone for each method. Similar NMDS and MRPP comparisons of fish species composition and common IBI metrics between upstream and downstream samples and among upstream samples only were also performed. For periphyton, total density and taxa richness were calculated and compared between upstream and down stream samples and among upstream samples using the nonparametric Mann-Whitney U test. Metrics were not used for periphyton analysis because currently there is no commonly accepted set of metrics for this assemblage.

Metrics Analysis. The multivariate approach has the disadvantage that it cannot determine if a particular difference in similarity represents an impairment or not. To address this issue, the project team also examined differences in actual metric values between samples to determine whether changes observed represent potential impairment for a given assemblage. In addition, examination of metrics and particular species, as well as physical habitat information were used to help identify whether observed changes in assemblage characteristics could be due to differences in flow or other habitat factors, as opposed to water quality changes.

By design, similar field and laboratory protocols were used across all facilities, therefore, it was not appropriate to rely on State-calibrated Indexes of Biological Integrity (IBI) for analyses. For both fish and macroinvertebrates, a suite of metrics commonly used by many states (Barbour et al. 1999, Karr and Chu 1999), was calculated, many of which are used by the states in which facilities were located (Table 2-6). For both fish and macroinvertebrates, 11 metrics were calculated addressing structural and functional aspects of the assemblages. To allow the 11 metrics to be aggregated to a single site score, metric values for each sample were converted to a 100-point scale using the formula

$$x * \frac{100}{A_{\max}}$$

where x is the calculated metric value, and A_{max} is the maximum value for a metric observed among all samples for a given facility. For reverse metrics, i.e., those that increase in the presence of stressors, the final value was derived as

Overall site scores (x_{tot}) were obtained by averaging the 11 scaled metric scores.

$$100 - \left(x * \frac{100}{A_{\max}} \right)$$

Performance of biological assessment protocols was evaluated using similarity analyses as well as documentation of their ability to attain the project MQO. Precision of field sampling was calculated using several parameters on sample data obtained from duplicate reaches upstream of the outfall at each facility, including relative percent difference (RPD), root mean squared error (RMSE), and coefficient of variation (CV). Calculations were based on individual metric values (x as calculated, not standardized), as well as x_{tot}. These calculations were performed for all three assemblages.

Relative percent difference (RPD) represents the proportional difference between two measures and is calculated using the equation:

$$RPD = \left(\frac{|A - B|}{A + B} \times 2 \right) \times 100$$

where, A is the metric or index value of the first sample and B is the metric or index value of the second sample (Berger et al. 1996).

Also called standard error of estimate, the root mean squared error (RMSE) is an estimate of the standard deviation of a population of observations and is calculated by:

$$RMSE = \sqrt{\frac{\sum_{j=1}^{k} \sum_{i=1}^{n_j} (y_{ij} - \bar{y}_j)^2}{\sum df_{1\ldots k}}}$$

where y_{ij} is the ith individual observation in group j, j = 1…k (Zar 1999).

Coefficient of variation (CV) is a unit-less measure calculated from the RMSE as:

$$CV = \frac{RMSE}{\bar{Y}} \times 100$$

where $\bar{Y}$ is the mean of the dependent variable (e.g., metric, index; Zar 1999). It is also known as relative standard deviation (RSD).

Differences between up and downstream samples for each assemblage and each facility were examined to help recognize associations with WET results and in other analyses relating WET and bioassessment results as described in the next section.

Relating WET and Bioassessment Results. Relationships between WET and biological assessment results were analyzed for each WET test and biological assessment type comparison (nine comparisons in all—three WET test types and three biological assessment types) (Figure 2-4). As described under Section 2.1 above, each WET test endpoint has a percent "fail" rate (i.e., effects were observed at test concentrations at or below facility's in-stream waste concentration). These fail rates were compared with each biological assessment metric "impaired" rate to determine the extent of overlap. "Impaired" status was assigned if NMDS and MRPP analyses indicated a significant change in a metric or species composition between up and downstream samples and the change was ecologically meaningful based on metric analyses; e.g., a decrease in a pollution-sensitive metric downstream as compared to upstream. For macroinvertebrate assessments, WET results were compared with results from both types of macroinvertebrate protocols (dip-net and Helter Dendy). In addition, the project team examined the degree to which each WET test type result was consistent with its "corresponding" biological assessment type: *Selenastrum* with periphyton biological assessment, *Ceriodaphnia* with macroinvertebrate assessment, and fathead minnow with the fish assessment. These analyses used the "fail" rate based on both design conditions and actual dilution conditions.

Table 2-1. Summary of Participating Facilities Including Influent/Effluent Characteristics, Treatment Plant Characteristics, as well as Previous Toxicity and WET Test Frequency.
MGD = Millions of Gallons Per Day; IWC = Instream Waste Concentration Under Design Low Flow Conditions.
Cd = *Ceriodaphnia dubia*, Pp = *Pimephales promelas*

Facility Code	EPA Region	Type of Facility (Industry or Municipal)	Design Flow (MGD)	Design IWC (%)	Hydraulic Retention Time (h)	Chronic Toxicity in 12 months previous to project	WET Frequency	WET Species
1	8	Municipal	36.3	62.3	12	No	Quarterly	Cd; Pp
2	8	Municipal	185	98.6	12	No	Monthly	Cd; Pp
3	2	Municipal	2.5	93	87.6	Yes* (Cd)	Quarterly	Cd
4	3	Municipal	1.5	86.5	16	No	Quarterly**	Cd; Pp
5	4	Municipal	6.2	97	Not reported	Yes (Cd)	Quarterly	Cd
6	4	Industrial	3.0	100	Not reported	No	Quarterly	Cd; Pp

* Reported toxicity was due to lab error; facility has been in compliance since this time.
** Conducted only upon permit renewal, every 5 years.

Table 2-2. Summary of Test Conditions for Three-Brood *Ceriodaphnia dubia* Survival and Reproduction Tests

Mean Temperature:	25 ± 1°C
Light quality:	Ambient laboratory illumination
Light intensity:	50-100 fc
Photoperiod:	16 hours light/8 hours dark
Size of test vessel:	60 mL
Volume of test solution:	30 mL
Renewal of test solution:	Daily
Age of test animals:	< 24 hours; all released within an 8 hour period
Number of animals per test vessel:	1
Number of replicate test vessels per concentration:	10
Total number of organisms per concentration:	10
Feeding regime:	Feed 0.10 mL of YTC and algal suspension per test chamber daily.
Aeration:	None
Control/dilution Water:	Moderately hard reconstituted water
Effluent concentrations:	Effluents: 5 and a control (required minimum) Receiving water: 100% receiving water and a control (recommended)
Test duration:	Until 60% of control females have three broods.
Effect measured:	Survival and reproduction
Test acceptability:	At least 80% survival; an average of at least 15 young/surviving female in the control solutions, and 60% of control females have three broods.

Table 2-3. Summary of Test Conditions for *Pimephales Promelas* 7-Day Chronic Survival and Growth Test.

Mean Temperature:	25 ± 1°C
Light quality:	Ambient laboratory illumination
Light intensity:	50-100 fc
Photoperiod:	16 hours light/8 hours dark
Size of test vessel:	473 mL
Volume of test solution:	250 mL
Renewal of test solution:	Once daily.
Age of test animals:	<24 hrs
Number of animals per test vessel:	10
Number of replicate test vessels per concentration:	4
Total number of organisms per concentration:	40
Feeding regime:	Feed *Artemia nauplii* a minimum of 2 hours prior to use in test, and twice daily during test, at least 6 hours apart.
Aeration:	As needed if D.O. in any test chamber falls below 4.0 mg/L
Control/dilution Water:	Moderately hard reconstituted water
Test concentrations:	Effluents: 5 and a control (required minimum) Receiving water: 100% receiving water and a control (recommended)
Test duration:	7 days
Effect measured:	Mortality or immobility, growth, biomass
Test acceptability:	At least 80% survival in controls; 0.25 mg dry weight per individual in controls

Table 2-4. Summary of Test Conditions for 96 Hour *Selenastrum capricornutum* Growth Toxicity Tests.

Mean Temperature:	25 ± 1°C
Light quality:	"Cool white" fluorescent lighting (recommended)
Light intensity:	86 ± 8.6 $\mu E/m^2/s$ (400 ± 40 ft-c or 4,306 lux) (recommended)
Photoperiod:	Continuous illumination (required)
Size of test vessel:	125 mL or 250 mL (recommended)
Volume of test solution:	50 mL or 100 mL[2] (recommended)
Renewal of test solution:	None (required)
Age of algal solution:	4 -7 days (required)
Initial cell density in test chambers:	10,000 cells/mL (recommended)
Number of replicate test chambers per concentration:	4 (required minimum)
Shaking rate:	100 cpm continuous, or twice daily by hand (recommended)
Aeration:	None (recommended)
Control/dilution Water:	Algal stock culture medium, enriched uncontaminated source of receiving or other natural water, synthetic water prepared using MILLIPORE MILLI-Q® or equivalent deionized water and reagent grade chemicals, or DMW
Test concentrations:	Effluents: 5 and a control (required minimum) Receiving water: 100% receiving water and a control (recommended)
Test duration:	96h (required)
Endpoint:	Growth (cell counts, chlorophyll fluorescence, absorbance, or biomass) (required)
Test acceptability:	Mean cell density of at least 1 x 10^6 cells/mL in the controls; and variability (CV%) among control replicates less than or equal to 20% (required)

Table 2-5. Whole Effluent Toxicity (WET) and Quality Control Toxicity Testing Performed by Each Facility.
"-" denotes that a facility did not conduct the test; "blank" denotes that the laboratory conducted the analysis for multiple facilities.

Facility	1st Quarter			2nd Quarter			3rd Quarter			4th Quarter			Blank 1			Blank 2			Duplicate			Split		
	Pp	*Cd*	*Sc*	*Pp*	*Cd*	*Sc*	*Pp*	*Cd*	*Sc*	*Pp*	*Cd*	*Sc*	*Pp*	*Cd*	*Sc*	*Pp*	*Cd*	*Sc*	*Pp*	*Cd*	*Sc*	*Pp*	*Cd*	*Sc*
1	√	√	-	√	√	-	√	√	√	√	√	√	√	√		√	√		√	√	-	√	√	√
2	√	√	-	√	√	√	√	√	√	√	√	√	√	√		√	√		√	√	-	√	√	√
3	√	√	-	√	√	√	√	√	√	√	√	√	-	-	√	√	√	√	-	-	√	√	√	√
4	√	√	-	√	√	√	√	√	√	√	√	√	√	√		√	√		-	-	-	√	√	√
5	-	√	-	-	√	√	√	√	√	√	√	√	√	√		√	√		√	-	-	√	√	√
6	√	√	√	√	√	√	√	√	√	√	√	√	√	√	√	√	√	√	√	√	√	√	√	√

Table 2-6. Benthic Macroinvertebrate Metrics Used for Analyses in this Project, and Their Use by Different States in Routine Assessment Programs. "Response to Stressors" Indicates Expected Direction of Change When Stressors are Present.

Metrics	Expected response to stressors	States where facilities located (Facility #)				
		CO (1,2)	NJ (3)	NC (5)	PA (4)	TN (6)
Total Taxa	decrease		x	x	x	
EPT Taxa	decrease	x		x	x	x
EPT Percent	decrease					x
Ephemeroptera Percent	decrease					
Intolerant Taxa	increase					
Tolerant Percent	decrease					
Dominant One Percent	decrease					
Filterers Percent	increase		x			
Scraper Percent	increase					
Clinger Taxa	decrease					
Clinger Percent	decrease					x

Figure 2-1. Map of Participating Facilities

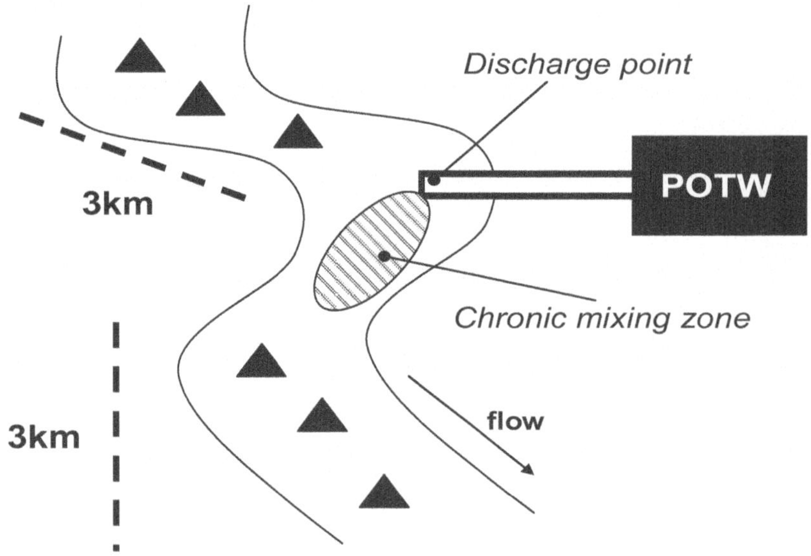

Figure 2-2. Biological Stream Sampling Design; Triangles Indicate Sampling Locations Upstream and Downstream of the Outfall From a Publicly Owned Treatment Works (POTW).

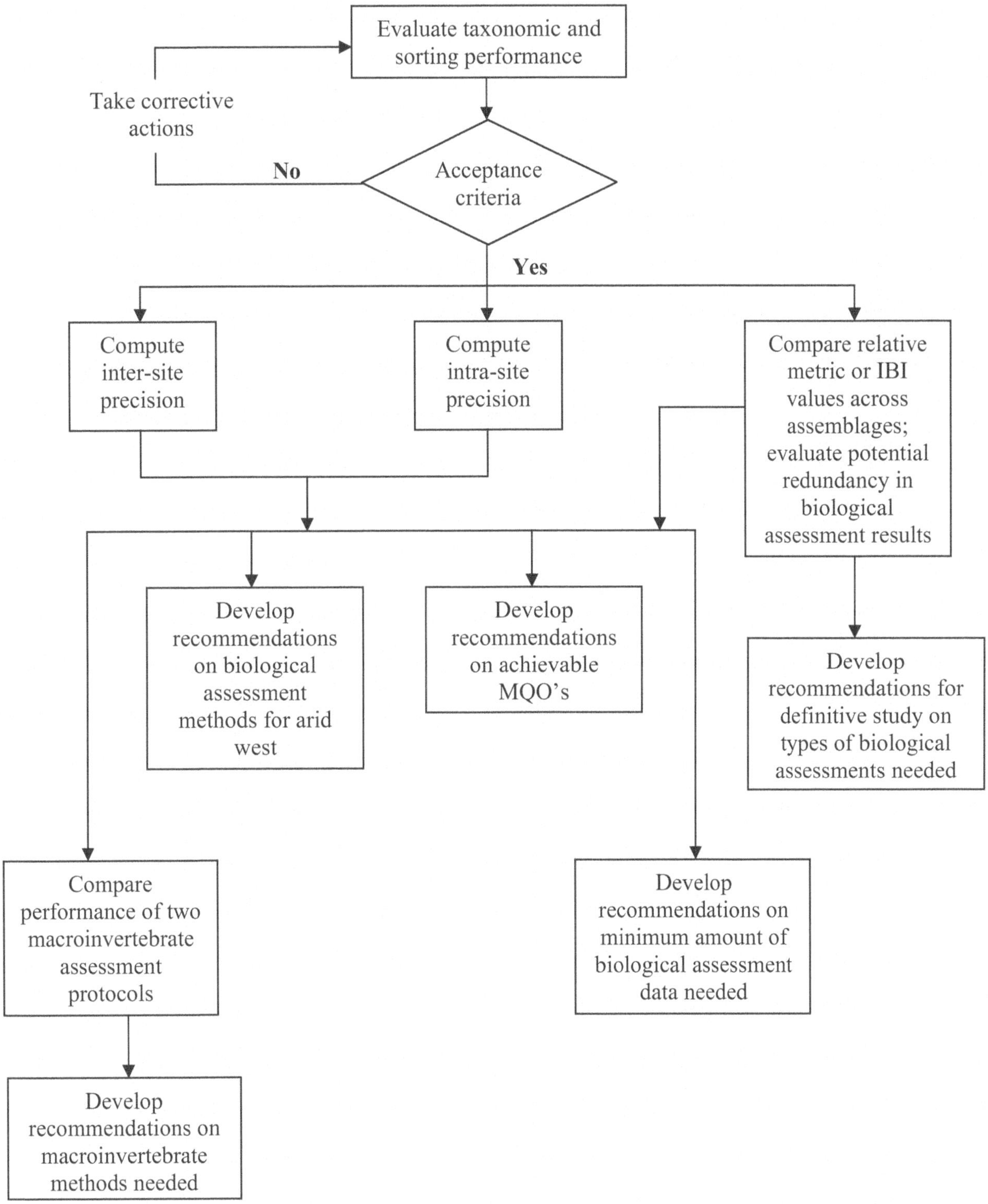

Figure 2-3. Flowchart Summarizing Analysis Tasks for Biological Assessment Results and Their Relationship to Study Objectives.

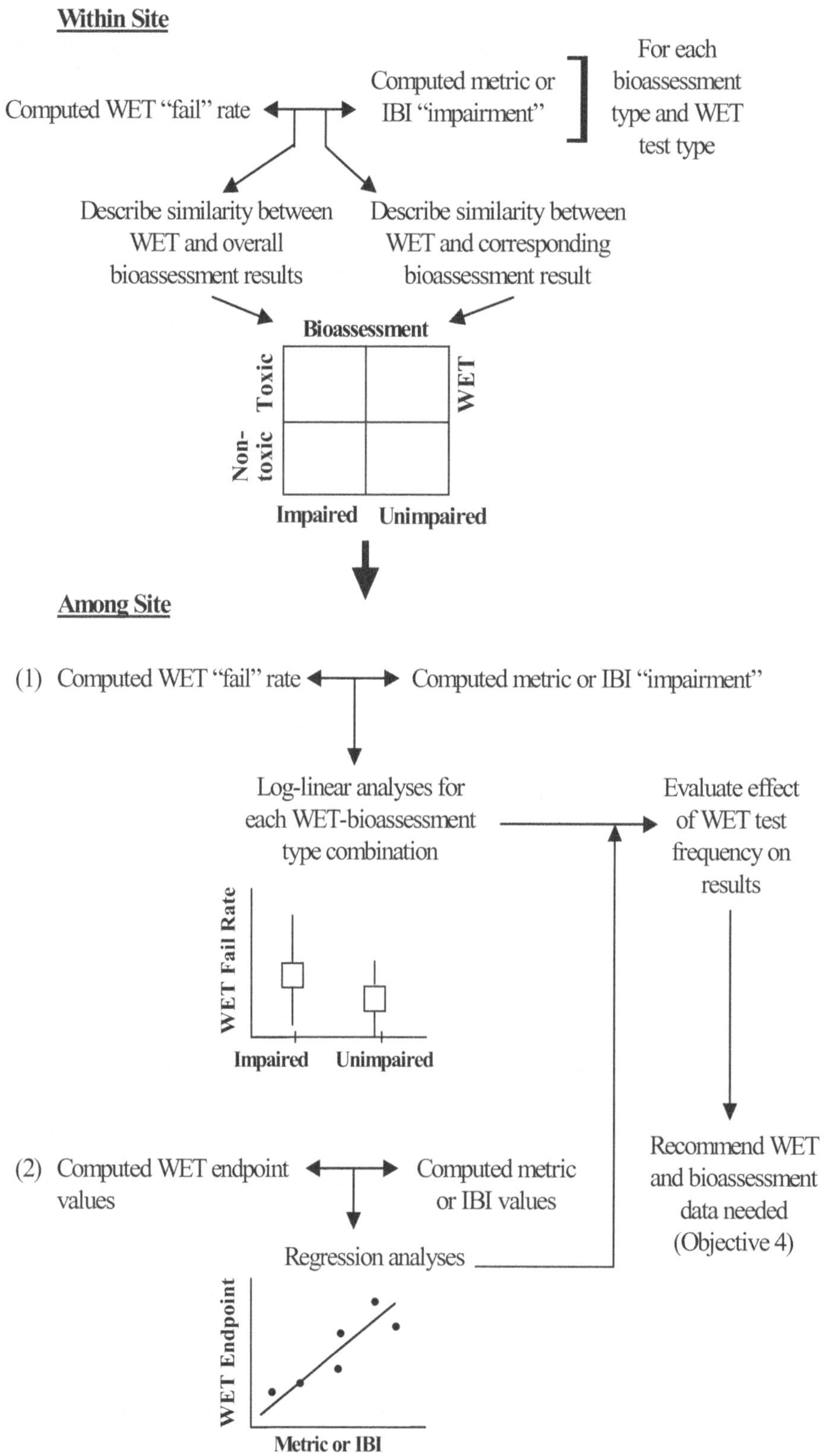

Figure 2-4. Flowchart Summarizing Analysis Tasks Relating WET to Biological Assessment Results and Relationship with Study Objectives.

CHAPTER 3.0

WET RESULTS

3.1 Compliance with WET Test Acceptance Criteria

The criteria evaluated in quality control (QC) reviews of WET tests were divided into two groups; test acceptability criteria (TAC) and test evaluation criteria (TEC). Tests not meeting any of the designated TAC (U.S. EPA, 2002a) were considered invalid. Those tests that failed to meet a single TEC should not be considered invalid, but failure to meet multiple TEC in a single test could be a concern.

3.1.1 *Ceriodaphnia dubia*

Of 35 *C. dubia* tests conducted, 97.1% met all TAC (Table 3-1 and Appendix D, available by request, Table D-1). Surprisingly, 37.1% of test reports included arithmetic errors in the calculation of reproduction summaries. The test reports also frequently failed to include critical information on sample receipt temperature (lacking in 51.4% of reports) and period between sample collection and test initiation (lacking in 48.6% of reports). Overall, the vast majority of the tests were acceptable and problems were isolated to reporting and documentation.

3.1.2 Fathead Minnow

Of 36 fathead minnow tests conducted, only 5.6% failed to meet the control performance requirements (Table 3-2 and Appendix D Table D-2). Although fewer were identified than were observed in *C. dubia* test reports, arithmetic errors were present in 11.1% of tests performed. Also similar to the *C. dubia* reports was the lack of information regarding sample receipt temperatures (lacking in 44.4% of reports) and period between sample collection and test initiation (lacking in 38.9% of reports).

3.1.3 *Selenastrum capricornutum*

The *S. capricornutum* tests had the fewest QC issues of the three tests types (Table 3-3 and Appendix D, Table D-3). Of the 26 tests performed, 96.2% (25 of 26) were considered valid based upon TAC. The only other problems identified were temperature excursions in one test and failure to document test sample holding time in two reports (7.7% of total).

3.2 Reference Toxicant Test Results

Five of the six facilities conducted 7-day chronic reference toxicant tests for *C. dubia* and *P. promelas* (Appendix E, available by request). Facility 6 conducted reference toxicant testing using *S. capricornutum* because they routinely conducted those tests while the others did not. Facility 2 performed acute reference toxicant tests; their permit did not require chronic reference toxicant testing. Of those that do perform chronic reference toxicant tests, not all followed the same protocol. In total, three toxicants were used among the five testing facilities. Facility 5 uses nonphenyl ethyloxylate (NPE) for their *C. dubia* testing and Facility 3 uses potassium chloride for their *C. dubia* and *P. promelas* testing. All other facilities use sodium chloride as their reference toxicant (including Facility 6 for *S. capricornutum* testing). Another difference among laboratories was the frequency of reference toxicant tests. Facility 5 performs them on a quarterly basis while the remaining testing facilities do so on a monthly basis.

Figures 3-1 and 3-2 summarize reference toxicant test results by species and endpoint, as well as by NOEC or IC_{25} value.

Using NPE as the reference toxicant, Facility 5 reported coefficients of variation (CV) for survival and reproduction of 7.6 and 14.9%, respectively, which met the desired MQO $\leq$ 35%. These CVs were less than those observed for *C. dubia* reference toxicant tests using sodium chloride, but comparable to those using potassium chloride.

Facility 3, which uses potassium chloride as a reference toxicant, reported CVs for *C. dubia* survival and reproduction of 19.9 and 15.3 %, respectively, also meeting the desired MQO. This facility reported that *P. promelas* was almost as sensitive to potassium chloride as *C. dubia*. C.V.s for fish survival and growth using potassium chloride were 10.1 and 9.6%.

The three laboratories using sodium chloride had similar estimates (Figure 3-2) for the 7-d *C. dubia* survival IC_{25}. C.V.s for survival ranged from 7.6 to 36%, with most of them meeting the MQO of CV $\leq$ 35%. This was also true for *P. promelas*, CV for fish survival ranged between 10.1 and 54.9%, with one laboratory (N=4) not meeting the MQO in general. The mean fish IC_{25} value from Facility 4 was significantly different from that reported for Facilities 5 and 6 (Figure 3-2). This indicates that fish mortality response to sodium chloride is not uniform among testing laboratories. Similar findings were made when considering lethal 7-day NOECs (Figure 3-1).

Sub-lethal endpoint values for *C. dubia* were more variable within laboratories than survival endpoints using sodium chloride. While all facilities reported similar *C. dubia* reproduction endpoints, CVs ranged from 19.9-68.3%, with two facilities exceeding the MQO (Figures 3-1 and 3-2). CVs for fish growth ranged between 8.9 and 32% indicating that laboratories could meet the MQO for this endpoint. Results from Facilities 1 and 6 were different than Facility 5 with respect to sub-lethal fish growth IC_{25} values. No difference was observed among the fish sublethal growth NOEC values.

All facilities using sodium chloride had some difficulty meeting the MQO for inter-test variability in reference toxicant tests (Table 3-4). Results of this analysis indicate that at least two laboratories could meet the MQO $\geq$ 90% of the time for each test endpoint in reference toxicant testing but the same lab could not always meet the MQO for all endpoints. Out of the five laboratories for which long-term chronic reference toxicant data were available, four labs met an MQO of $\leq$ 35% difference in endpoint value at least 75% of the time (9 out of 12 monthly tests) with the exception of the *C. dubia* reproduction endpoint. For that endpoint, a greater amount of inter-test variability in endpoint value was observed than for other endpoints and an MQO of < 50% difference at least 75% of the time appeared to be feasible for the laboratories in this study.

3.3 Effluent Test Results

WET testing conducted for this project, in large part, exhibited a low level of chronic toxicity using either *C. dubia* or *P. promelas* (Figure 3-3 and 3-4; Appendix F, available by request). Effects, if observed, were generally related to sublethal endpoints as opposed to survival for these species. *P. promelas* test results consistently yielded NOEC values lower than their respective IC_{25} values indicating differences in results depending on the statistical method used. For the most part, facilities exhibited a fairly limited range of endpoint values over the four quarters of testing for both of these test types. However, selected tests from some facilities exhibited a fairly broad range of endpoint values over the year of testing (e.g., Facility 3 *S. capricornutum* IC_{25} or Facility 6 *P. promelas* growth NOEC).

S. capricornutum tests were conducted in three rather than four quarters as planned for most facilities as pointed out in Section 2.4 due to the fact that monitoring in this project started after the first quarter of testing for most facilities. The *S. capricornutum* test generally yielded the most sensitive WET results for the effluents tested in this project (Figure 3-4). NOEC and corresponding IC_{25} values tended to be relatively low (often < 20% or > 6 TU_c for many samples examined, suggesting a high level of toxicity to this species). One exception to this observation occurred at Facility 6 where no toxicity was observed to this species.

3.4 Evaluation of WET Test Precision

In addition to meeting test acceptance criteria, this study also evaluated test performance in terms of meeting proposed MQOs based on Percent Minimum Significant Differences (PMSDs) calculated for each effluent test. This analysis focused on sublethal endpoints and proposed MQOs based on maximum PMSDs for each test: reproduction for *C. dubia* (PMSD <37%), and growth for *S. capricornutum* (PMSD <35%) and *P. promelas* (PMSD <35%). These PMSD levels were recommended in EPA's WET variability guidance (U.S. EPA, 2002b) and their inter-laboratory study (U.S. EPA, 2000b). Those PMSD values were based on 90th percentiles from tests using *P. promelas* (19 laboratories, 205 tests), *C. dubia* (33 laboratories, 393 tests), and *S. capricornutum* (9 laboratories, 85 tests) chronic WET tests. In comparison, this pilot study was much smaller in size, so the PMSDs observed in this study might not be representative of what WET laboratories are capable of consistently achieving in general.

For the *S. capricornutum* tests, the 90th percentile PMSD was 32.0%, which met the proposed MQO (< 35%) (Figure 3-5) . In fact, all of the 19 algal WET tests conducted in this study produced a PMSD less than the proposed MQO indicating satisfactory test precision for this method. For the *P. promelas* test, the 90th percentile PMSD was 27.2%, which also met its MQO (< 35%; Figure 3-6). Out of 31 *P. promelas* tests, 3% (1 test) exceeded the MQO and the highest PMSD observed in this study was 43.8%. Many laboratories did not meet the MQO in the *C. dubia* test (MSD < 37%; Figure 3-7) in this study. Out of 33 *C. dubia* WET tests conducted in this study, PMSDs based on reproduction did not meet the MQO in 24%, or 9, of the tests. The highest PMSD observed for this test was roughly twice the proposed MQO (PMSD = 78.4%). The 90th percentile PMSD for this test was 53.2%. Thus, *C. dubia* test results were much more variable in this pilot study than they were in U.S. EPA's inter-laboratory study.

3.5 Evaluation of False Positives

One of the blind samples tested by each laboratory in this study consisted of moderately hard reconstituted water (U.S. EPA, 2002a) prepared by the project team and distributed to each facility, which served as a negative control or blank sample. This sample was treated by the laboratories as an effluent sample as described in Section 2.2. Results of this testing indicate the extent of false positives for each of the three WET test methods used in this study.

All facilities reported no significant effects on survival or sub-lethal effects for *C. dubia* (Table 3-5; Appendix G, available by request). Two facilities reported adverse sublethal effects in *P. promelas* testing: these facilities reported growth NOEC values of 50 and <4%, respectively, although all facilities reported IC_{25} values >100% for this sample (Table 3-5). Both laboratories that conducted the *S. capricornutum* test reported adverse effects due to the blank sample with IC_{25} values of 53 and 57% and NOECs equal to 40% sample (Table 3-5).

3.6 Evaluation of False Negatives

A second blind sample, tested by each laboratory, was used to examine false negatives for each WET test. This test used a blind sample containing 6.4 mg/L copper which was then diluted by each laboratory according to the testing protocol provided for each species. Thus, the labs treated the sample as if it was a normal effluent sample in testing as described in Section 2.2. The dilution regime used in the *C. dubia* test was designed to produce a toxic response only at higher sample concentrations (test sample copper concentration was 16 μg/L). For *P. promelas* and *S. capricornutum*, the dilution instructions were designed to produce a toxic response at mid-concentrations of sample. Hardness was not a significant factor in these tests because all participating laboratories used moderately hard water (80-100 mg/L as $CaCO_3$).

Endpoint values reported for these tests were fairly consistent among laboratories for *P. promelas* and *S. capricornutum*, while *C. dubia* endpoints were more variable among laboratories (Figures 3-8 and 3-9; Appendix G). Sub-lethal 7-day NOEC values for *C. dubia* and *P. promelas* displayed similar variability among laboratories as the survival endpoints (Figures 3-8 and 3-9).

IC_{25} and NOEC values reported in *C. dubia* tests were not consistent among laboratories for the positive control sample. Two facilities reported survival IC_{25} and NOEC values < 63 % sample while four facilities reported little or no significant lethality. Half (3) of the facilities observed sub-lethal effects for *C. dubia*, all at sample concentrations ≤ 67% sample. The other three facilities reported no sublethal effects of this sample to *C. dubia*. Of the three facilities that did report significant effects of the sample on *C. dubia*, two facilities reported sublethal endpoints equal to the survival endpoints (whether based on NOEC or IC_{25} values).

In the fish tests, five of the six facilities reported IC_{25}s for survival ≤ 43.5% and all reported survival NOEC values ≤ 25% sample. For fish growth, all facilities reported IC_{25} values ≤ 25% sample, indicating consistency in the sample being recognized as toxic to *P. promelas*. In terms of the IC_{25} values reported for this test, all six facilities reported more sensitive endpoints for growth than for survival. For NOEC values, four of the six facilities reported more sensitive endpoints for growth than for survival.

Algae testing using this blind sample generated disparate endpoints between the two testing facilities. IC_{25} values differed by a factor of 9 while NOEC values differed by a factor of 4 (two test concentrations apart). Both IC_{25} and NOEC values were significantly different ($p < 0.01$) between the two laboratories using this blind sample.

3.7 Results of Split Sample Testing

Survival endpoints for *P. promelas* and *C. dubia* were consistent between laboratories based on either IC_{25} or NOEC values. In general, all of these effluent samples were fairly non-toxic to the test species in terms of survival (IC_{25} or NOEC = 100% effluent). Growth or reproduction endpoints for these two species, however, showed greater variability between laboratories in general. Testing with *S. capricornutum* indicated even greater variability between the two participating laboratories for the same effluent samples. A summary of interlaboratory results is presented below by species test (see Appendix H, available by request, for raw data).

3.7.1 *Pimephales promelas*

Values for the survival endpoints (IC_{25}, NOEC, LOEC) were very consistent between laboratories for each facility with the exception of Facility 1 (Table 3-6). On the other hand, growth endpoints (either IC_{25} or NOEC) were somewhat variable among laboratories. Some endpoints were different by as much as a factor of 8 (i.e. NOEC values obtained by the two laboratories for the same effluent samples from Facility 6 were 100 and 12.5%).

In general, both IC_{25} and NOECs were equally different between laboratories for this set of effluent samples.

3.7.2 *Ceriodaphnia dubia*

Results from split testing using *C. dubia* yielded more consistent results than those from the corresponding *P. promelas* tests (Table 3-7). Survival endpoints were relatively consistent between laboratories with the exception of two facilities (Facilities 3 and 6), however, these split test results were within 95% confidence intervals for the two endpoints.

Reproduction endpoints between laboratories were also relatively consistent in this set of tests with the exception of Facilities 2 and 6 in which significant differences in both IC_{25} and NOEC values were observed between laboratories. The maximum difference in IC_{25}s observed was a factor of about 4 (Facility 2) and averaged a factor of 1.27 for this set of tests. Maximum and average difference in NOECs between laboratories could not be determined because one laboratory NOEC was a "less than" value (Facility 2) but is at least a factor of 2 in this set of tests.

3.7.3 *Selenastrum capricornutum*

Results of split sample WET testing with *Selenastrum capricornutum* yielded high variability between the two laboratories in some cases (Table 3-8). Both IC_{25} and NOEC endpoints for Facilities 1, 3, 4, and 5 indicated different results between laboratories, with no toxicity observed in one laboratory and a high level of toxicity observed in the other laboratory. IC_{25} values between laboratories differed by as much as about 14 times in the case of Facility 3 and averaged 5.5 times difference for this set of tests. NOEC values differed by as much as 16 times (Facilities 2 and 5) and averaged 6.83 times different between laboratories for this set of tests. Consistent results were observed for this test for Facilities 2 and 6, with Facility 6 indicating no toxicity to this species by both laboratories.

3.7.4 Evaluation of Split Sample Results

As a way of displaying the potential significance of the split test results, the laboratory results for each facility were compared with the facility's WET limit or trigger concentration (based on design dilution instream) to determine whether a given laboratory result would be assessed as a "pass" (in compliance) or a "fail" (not in compliance). A comparison of the split testing is illustrated in Figure 3-10. These charts visually compare the number of times the two laboratories both reported results that would lead to the interpretation that a given test passes or fails, as well as whether their results lead to contradictory interpretations regarding effluent compliance. The analysis is grouped by species test (*C. dubia*, *P. promelas*, and *S. capricornutum*) and endpoint (IC_{25} or NOEC). The algal growth test generated the most contradictory results between laboratories. For both types of statistical endpoints (IC_{25} or NOEC), four out of six split tests yielded contradictory interpretations between the laboratories. For the fish 7-day WET tests, the two laboratories reported results leading to contradictory interpretations half of the time, based on either the IC_{25} or the NOEC. The *C. dubia* 7-day WET test yielded the most reproducible results in split sample testing, with only one out of six facility tests yielding contradictory interpretations using IC_{25} and two out of six yielding contradictory interpretations using the NOEC.

3.8 Results of Duplicate Testing

3.8.1 *Ceriodaphnia* and *P. Promelas*

Testing with *C. dubia* or *P. promelas* showed no intra-laboratory variability in endpoints among duplicate tests (Tables 3-9 and 3-10). Both NOECs and IC_{25}s were > 100% effluent for each facility in which duplicate tests were conducted.

3.8.2 *S. capricornutum*

Duplicate testing with *S. capricornutum* was completed for two facilities (Table 3-11). Endpoints for these tests were similar, one with identical endpoint values (Facility 6) and the other with values within the 95% confidence intervals of each value.

3.9 Prediction of Potential Effluent Effects Instream

In these analyses, WET test endpoints were evaluated to determine whether the results from a given test suggest that the effluent is unlikely to result in instream toxicity (i.e., test "passes") or could result in instream toxicity (i.e., test "fails"). In these analyses, the project team examined both the frequency and magnitude of WET excursions in order to determine potential for instream toxicity for each facility (Diamond et al., 1999; Diamond and Daley, 2000). Frequency of excursions is thought to be important because a facility that more frequently exceeds their WET limit or trigger concentration should have a higher likelihood of producing biological effects in the receiving stream, all other factors being equal. Magnitude of toxicity could also be important, even if the frequency is relatively low, because effects on biological assemblages could be large and recovery much slower.

WET test passes or fails were determined by comparing test endpoints to the facility's WET limit or trigger concentration. IC_{25}s and NOECs for both survival and sublethal endpoints were evaluated in these analyses. For all of the facilities in this study, the regulatory limit or trigger was based on design flow conditions. In addition, the project team evaluated test pass/fail on the basis of actual flow conditions at the time of testing. An implicit assumption in this analysis is that the WET test results observed are representative of each facility's actual WET condition over the year. Since most facilities conducted quarterly WET tests (with the exception of Facility 2 which conducted monthly tests), the degree of representativeness of test results and the calculation of the percentage of test "fails" appears justified.

C. dubia and *P. promelas* survival endpoints resulted in predominantly passing results for most facilities when based on actual effluent dilution conditions (Figure 3-11). This was especially true when IC_{25} endpoints were used. Comparisons with limits or trigger concentrations based on design effluent dilution produced more "failed" tests than when compared to actual dilution concentrations as expected. Sublethal endpoints resulted in more "failed" tests than survival endpoints, using either IC_{25} or NOEC values, especially when compared to limits based on design effluent dilution.

Based on typical design flow permit limits, facilities had proportionally the highest "failed" test rate based on the *Selenastrum* test, followed by the *P. promelas* test, and the *C. dubia* test (Table 3-12). The "failed" test rate was higher using the NOEC as compared to the IC_{25} for all three test types. These results indicate that at least in some cases, the NOEC corresponded to less than a 25% decrease in organism response from controls.

The results presented in Table 3-12 indicate that most facilities did not pass the *S. capricornutum* tests. Only Facility 6 passed these tests consistently. Sites 1, 2, 4, and 5 had no "passing" tests. The same testing laboratory conducted *S. capricornutum* tests for Facilities 1 through 5. Facility 6 used a different laboratory. Failure rates observed using *S. capricornutum* maybe suspect.

Given the *C. dubia* and *P. promelas* results above, Facility 3 had the highest overall percentage of "failed" WET tests based on design flow dilution, followed by Facilities 2, 1, 6, 4, and 5 in decreasing order. Based on actual dilution conditions, Facility 1 had the highest percentage of "failed" WET tests followed by Facilities 3, 2, 4, 6, and 5 in decreasing order.

As mentioned previously, another factor that was evaluated as an indicator of predicted effects instream in this analysis was magnitude of "failed" tests, expressed in terms of the facility WET limit or trigger. The project team hypothesized that the magnitude of effects on biota instream is related to the magnitude of apparent effluent toxicity.

Using magnitude of toxicity to *C. dubia* or *P. promelas* as an indicator, Facility 6 reported the most toxicity relative to its design flow trigger with a *P. promelas* test yielding an NOEC 16 times its design WET trigger (i.e., 16 TU_c). Facility 2 had the next highest magnitude toxicity, in which two *C. dubia* test IC_{25}s were 13 and 15 times its design WET trigger (i.e., 13 and 15 TU_c). Highest magnitude of toxicity observed for the four other facilities in this study were considerably lower (1.1-1.9 TU_c, based on NOEC values). The greatest potential for instream effects, based on the highest degree of toxicity exhibited in WET tests and design flow conditions in decreasing order were Facility 6, 2, 3, 4, 1, and 5.

If magnitude of toxicity was evaluated in terms of actual effluent dilution at the time of testing, then Facility 6 still exhibited the most toxicity (16 TU_c) followed by Facilities 2 (2.4 TU_c), and 3 (1.14 TU_c). The remaining facilities all had *C. dubia* and *P. promelas* endpoint TU_cs < 1.0 based on actual dilution.

Including results of *S. capricornutum* in this analysis changes some of the results described above. The highest toxicity observed based on *S. capricornutum* IC_{25} or NOEC values and design dilution was for Facilities 5 (15.5 TU_c) and 3 (14.9 TU_c), followed by Facilities 1 (10 TU_c), 2 (8 TU_c), 4 (3 TU_c), and 6 (1.0 TU_c). Based on actual dilutions, the highest algal toxicity was observed for Facility 5 (14.4 TU_c), followed by Facilities 2 (3.8 TU_c), 3 (3.7 TU_c), 4 (3.5

TU_c), 1 (1.4 TU_c), and 6 (1.0 TU_c) in decreasing order of toxicity. If the *S. capricornutum* WET results are accurate representations of potential effects on instream periphyton, these results predict greatest effects on the downstream periphyton assemblage at Facility 5, followed by 2, 3, and 4 in decreasing order, with little or no effects on periphyton predicted downstream of Facilities 1 or 6.

3.10 Summary of WET Results

The foregoing analyses of WET tests and associated quality control analyses conducted in this pilot study suggest several findings of relevance to the design of a future study that examines quantitative relationships between WET test results and instream biological condition. Some of these findings are process-related as opposed to technical but are equally important in terms of how WET testing should be conducted in such a study and the types of MQOs that can reasonably be required. It is important to note that this was a pilot study with relatively few facilities and laboratories participating. Therefore, one must be cautious in extrapolating the findings observed here to facilities and laboratories in general. The following section highlights the key WET findings based on results presented in this chapter.

3.10.1 Process-Related Findings

Reference Toxicant Testing Programs Vary Across Laboratories. As pointed out in Section 3.2, three different reference toxicants were used by the six laboratories involved in this Pilot. Furthermore, the frequency with which reference toxicant testing is conducted varied across laboratories from monthly to quarterly. In one case, the facility conducts acute reference toxicant testing but is required to conduct chronic WET tests monthly. The limited results here suggest that sodium chloride, the most common toxicant used in this study and by most WET labs, may not yield desired endpoint precision among reference toxicant tests as compared to potassium chloride or NPE. Furthermore, there appeared to be more variability in *P. promelas* endpoints among laboratories that use sodium chloride than for *C. dubia*, which may indicate subtle differences in fish batch sensitivity to this chemical or factors related to tests conducted with commercially supplied versus laboratory-cultured organisms. Reference toxicant tests did not always coincide with effluent testing.

Effluent Test Reporting and Documentation Varies Widely. The project team observed a fairly wide range in documentation detail among laboratories in their WET test reports. As many of these reports were copies of what was sent to the regulatory authority, it appears that these differences may be important. Key issues noted were: 1) arithmetic errors in calculating *C. dubia* reproduction or fish weight and number of broods per female in controls over the test, and 2) lack of information on sample holding times or sample temperature upon check-in. While most of the arithmetic errors were inconsequential in terms of the end results, it may be cause for concern if more widespread among WET testing labs.

3.10.2 Scientific Findings

A summary of the extent to which the various MQOs were met for WET testing in this Pilot is presented in Table 3- 13. Major findings are summarized below.

Test Acceptance Criteria Generally Met. Most labs met the test acceptance criteria for all tests conducted in this pilot. In two cases where invalid tests were reported by a laboratory, the test was re-run in the same quarter.

Reference Toxicant Endpoint Precision Between Tests Generally Met MQO. The variability in test endpoints between reference toxicant tests conducted over a year was often within the MQO of 35% of the long-term average value for different labs, but many tests could not meet this MQO. None of the labs could meet this MQO all of the time for all test types and it appears that this MQO is substantially more stringent than the control limits typically used by labs and recommended by U.S. EPA. Sublethal endpoints appeared to be more variable across tests than survival endpoints as expected because acceptance criteria for survival are fairly narrowly defined for each method. Most of the participating labs could meet an MQO of inter-test variability of 50% most (75%) of the time. It is not clear how much the inter-test variability observed in reference toxicant tests in this study is representative of the variability that might be observed within a lab using effluent samples because it is not feasible to repeatedly test the same effluent sample over the course of a year (because the sample holding time for the methods would be exceeded). However, it may be useful to at least document the degree to which a given reference toxicant test endpoint varies from the lab's long-term average value so as to provide another measure of test variability.

Effluent Test Precision and Sensitivity Generally Met MQOs. Most test controls exhibited satisfactory replication such that the recommended MQOs were met by most labs. This was particularly true for the *P. promelas* and *S. capricornutum* tests. The *C. dubia* tests displayed somewhat higher PMSDs with 76% of the *C. dubia* effluent tests reporting PMSDs < 40% (MQO was ≤ 37%). There was not necessarily a relationship between the labs that met the MQO for inter-test precision in reference toxicant tests and the MQOs for effluent test precision. For the *C. dubia* tests conducted in this project, an MQO of PMSD < 50% is very feasible (approximately 90% of all the tests conducted met this MQO).

Intra-Laboratory Precision of Effluent Tests is High. Duplicate testing in each laboratory produced repeatable test results. However, nearly all samples conducted in duplicate testing by each Facility happened to be non-toxic to both *C. dubia* and *P. promelas* (NOECs and IC_{25}s ≥ 100% effluent). This study cannot address whether similar replication of results for these species tests is expected with effluents that have more inherent toxicity. The *S. capricornutum* test was duplicated in only two facilities; one with a toxic effluent and one with a non-toxic effluent, and both sets of tests exhibited fairly high agreement.

False Positives Observed in Some Cases. Of the chronic tests conducted in false positive testing, two *P. promelas* tests (N=12) and both the *S. capricornutum* tests (N=2) resulted in reported toxicity. However, the *P. promelas* false positives were based on NOEC values not IC_{25}s. The toxicity observed using the algae test and reconstituted moderately hard water in this study was one of several issues observed with results using this test method.

False Negatives Observed in Some Cases. Out of the 14 tests conducted in this analysis, there were a few that reported no toxicity in the sample using the *C. dubia* test in which the toxic concentration was designed to be at the high end of the test concentration series (75-90% of the sample). Examining both lethal and sublethal endpoints for each test, no false negatives were observed based on either the fish or the algae test, in which a toxic concentration was designed to be in the low to mid-range of the test concentration series. For the *C. dubia* test, half of the six labs reported the sample as non-toxic (NOEC or IC_{25} = 100%) and half the labs reported the sample as toxic (NOEC = 25-50% or IC_{25} 30-65%). If the *C. dubia* results observed in this project are representative of WET labs in general, these analyses suggest that data quality issues may be a significant factor affecting WET pass/fail interpretations if the effluent is toxic only at higher concentrations. These results, as well as the split sample test results, suggest that laboratories using the same methods to test the same sample may disagree more often then expected.

Precision of Split Test Endpoints Often Less than Desired. The MQO for this analysis was < 35% difference in endpoints between laboratories for the split sample tests but this was not met in many of the tests, particularly for sublethal endpoints. The algae test resulted in the poorest agreement between the two labs and *C. dubia* had the best agreement. Generally, IC_{25} values between labs were within 50% of each other while NOEC values were often within 60% of each other. The variability in endpoint values reported in split testing could translate to differences between passing and failing a WET test in some cases.

NOEC Values More Conservative than IC25s. In general, effluent tests reported more toxicity using the NOEC as compared to the IC_{25} in this study. In several cases, the NOEC indicated a statistically significant effect at effluent concentrations well below 100% while the corresponding IC_{25} was >100%. When combined with design (i.e., low flow) effluent dilution as the compliance concentration in WET testing, NOEC endpoints often yielded more "failed" tests than IC_{25} endpoints, particularly based on sublethal effects.

S. capricornutum Test Yielded Most Sensitive Results. The green algae test resulted in the most toxicity for five of the six effluents examined in this project, with either NOECs or IC_{25}s well below those reported using either *C. dubia* or *P. promelas*. However, given the poor agreement observed in split testing, as well as the false positives observed using control laboratory water with this test, effluent test results using the algae test are considered suspect, pending further evaluation.

Compliance with Test Acceptance Criteria is only Part of the Story. All of the QC testing in this pilot suggests that compliance with test acceptance criteria, while necessary and important, may not be sufficient for evaluating laboratory performance and data quality for these WET tests. Other quality control analyses (many of which are recommended by U.S. EPA and other regulatory agencies) are needed to ensure that correct test interpretations are made.

Table 3-1. Summary of Laboratory Compliance with Test Acceptability and Test Evaluation Criteria for the *Ceriodaphnia Dubia* 7-Day Chronic WET Test.

Chronic *Ceriodaphnia dubia*

	Test Acceptability Criteria		Test Evaluation Criteria					
	Control Survival >80%?	Control Reproduction: >60% with 3 Broods with mean of 15 neonates?	Test Ended after controls had 60% reproduction?	Arithmetic errors on reproduction summaries?	Mean test Temperature 25 +/- 1 degree C?	Instantaneous test temperature always within 3 degrees C of goal?	Receipt Temperature from 0-6 degrees C?	Test initiated within 36 hours of sample collection?
Percentage of tests with deficient report for criterion	0.0%	0.0%	0.0%	0.0%	2.9%	2.9%	51.4%	48.6%
Percentage of tests compliant with Criterion	**97.1%**	**97.1%**	**91.4%**	**37.1%**	**97.1%**	**97.1%**	**45.7%**	**51.4%**

Table 3-2. Summary of Laboratory Compliance with Test Acceptability and Test Evaluation Criteria for the *Pimephales promelas* 7-day Chronic WET Test.

Chronic *Pimephales promelas* (fathead minnow)									
	Test Acceptability Criteria		Test Evaluation Criteria						
	Control Survival >80% ?	Mean dry mass of surviving control organisms > 0.25 mg ?	Arithmetic errors on mass summaries?	Test run for 7 days?	Mean test Temperature 25 +/- 1 degree C?	Instantaneous test temperature always within 3 ° C of goal?	D.O. always > 4.0 mg/L? (if not, was it aerated?)	Receipt Temperature from 0-6° C?	Test initiated within 36 hours of sample collection?
Percentage of tests with deficient report for criterion	0.0%	0.0%	0.0%	0.0%	0.0%	0.0%	0.0%	44.4%	38.9%
Percentage of tests compliant with Criterion	94.4%	100.0%	11.1%	100.0%	100.0%	100.0%	100.0%	50.0%	61.1%

Table 3-3. Summary of Laboratory Compliance with Test Acceptability and Test Evaluation Criteria for the *Selenastrum capricornutum* 4-day Chronic WET Test.

Chronic *Selenastrum capricornutum*

	Test Acceptablity Criteria		Test Evaluation Criteria							
	Final mean cell density of at least $1x10^6$ cells/ml in controls ?	CV among control replicates less than or equal to 20% ?	4 Reps per concentration; initial cell density of 10,000 cells/ml ?	Arithmetic errors on cell density summaries?	Chambers exposed to continuous light of 400 +/- 40 ft-c?	Were stock nutrients (with EDTA) added to each chamber?	Mean test T of 25 +/- 1 degree C?	Test T always within 3 degrees C of goal?	Receipt Temp of 0-6°C	Test initiated within 36 hours of sample collection?
Percentage of tests with deficient report for criterion	0.0%	0.0%	0.0%	0.0%	0.0%	0.0%	0.0%	0.0%	0.0%	7.7%
Percentage of tests compliant with Criterion	100.0%	96.2%	100.0%	100.0%	100.0%	100.0%	96.2%	100.0%	100.0%	92.3%

Table 3-4. Percentage of Chronic Reference Toxicant Test that Met the Suggested Measurement Quality Objective (MQO) of Being Within 35% of the Long-Term Average Endpoint Value (Either Survival or Sublethal Endpoint) for a Given Facility (Laboratory) and Test Type. All of the Laboratories Conduct Monthly Reference Toxicant Tests Except Facility 5 Which Conducts Them Quarterly. Facility 2 Does Not Conduct Chronic Reference Toxicant Tests and is Not Included.

Facility	C. dubia		P. promelas		S. capricornutum
	Survival	Reproduction	Survival	Growth	Growth
1	83	50	92	92	NA
3	92	100	100	92	NA
4	92	8	75	83	NA
5	100	100	44	74	NA
6	92	92	84	100	71.5

Table 3-5. Results of Blind Sample False Positive WET Testing for Each Facility. IC_{25} = Effluent Concentration Causing At Least a 25% Reduction Compared To Control Response; NOEC = No Observed Effect Concentration; LOEC = Lowest Observed Effect Concentration.

Facility	Test Species	Lethal			Sub-Lethal		
		IC_{25} (%)	NOEC (%)	LOEC (%)	IC_{25} (%)	NOEC (%)	LOEC (%)
1	*Ceriodaphnia dubia*	>100	100	>100	>100	100	>100
	Pimephales promelas	>100	100	>100	>100	100	>100
	*Selenastrum capricornutum**	NA	NA	NA	53.4	40	60
2	*Ceriodaphnia dubia*	>100	100	>100	>100	100	>100
	Pimephales promelas	>100	100	>100	>100	100	>100
	*Selenastrum capricornutum**	NA	NA	NA	53.4	40	60
3	*Ceriodaphnia dubia*	NM	NM	NM	NM	NM	NM
	Pimephales promelas	NM	NM	NM	NM	NM	NM
	*Selenastrum capricornutum**	NA	NA	NA	53.4	40	60
4	*Ceriodaphnia dubia*	>100	100	>100	>100	100	>100
	Pimephales promelas	>100	100	>100	>100	50	100
	*Selenastrum capricornutum**	NA	NA	NA	53.4	40	60
5	*Ceriodaphnia dubia*	>100	100	>100	100	100	>100
	Pimephales promelas	>100	100	>100	>100	100	>100
	*Selenastrum capricornutum**	NA	NA	NA	53.4	40	60
6	*Ceriodaphnia dubia*	>100	100	>100	>100	80	100
	Pimephales promelas	>100	100	>100	>100	<4	4
	Selenastrum capricornutum	NA	NA	NA	56.7	40	60

* - *S. capricornutum* testing for Facility 1-5 were performed by the same facility.

Table 3-6. Summary of Split WET Testing Using *Pimephales promelas* 7-Day Chronic Testing.

Facility Code	Lab	7-D Survival			7- D Growth		
		IC_{25} (%)	NOEC (%)	LOEC (%)	IC_{25} (%)	NOEC (%)	LOEC (%)
1	1	75	62.3	75	45.8	62.3	75
	2	>100	100	>100	>100	100	>100
2	1	>100	100	>100	50.8	25	50
	2	>100	100	>100	>100	100	>100
3	1	>100	50	100	>100	50	100
	2	>100	100	>100	>100	100	>100
4	1	>100	100	>100	72.4	100	>100
	2	>100	100	>100	>100	100	>100
5	1	>100	100	>100	>100	100	>100
	2	>100	100	>100	>100	100	>100
6	1	>100	100	>100	>100	100	>100
	2	>100	100	>100	38.4	12.5	25

Table 3-7. Summary of Split WET Test Results Using EPA's *Ceriodaphnia dubia* 7-Day Chronic Test.

Facility Code	Lab	Three-Brood Survival			Three-Brood Reproduction		
		IC_{25} (%)	NOEC (%)	LOEC (%)	IC_{25} (%)	NOEC (%)	LOEC (%)
1	1	>100	100	>100	>100	100	>100
	2	>100	100	>100	>100	100	>100
2	1	>100	100	>100	7.5	<12.5	12.5
	2	>100	100	>100	31.2	25	50
3	1	93.8	100	>100	97.8	100	>100
	2	>100	100	>100	>100	100	>100
4	1	>100	100	>100	>100	100	>100
	2	>100	100	>100	95.5	89	100
5	1	>100	100	>100	>100	100	>100
	2	>100	70	90	>100	100	>100
6	1	>100	100	>100	>100	100	>100
	2	82.2	100	>100	57.4	50	100

Table 3-8. Summary of Split Laboratory Test Results Using the EPA 4-Day *Selenastrum capricornutum* WET Test.

Facility Code	Lab	96-H Growth		
		IC_{25} (%)	NOEC (%)	LOEC (%)
1	1	21.4	50	62.3
	2	>100	100	>100
2	1	19.2	12.5	25
	2	11.2	<12.5	12.5
3	1	7.33	<6.25	6.25
	2	>100	100	>100
4	1	47.4	33	63
	2	93.4	100	>100
5	1	9.8	6.25	12.5
	2	>100	100	>100
6	1	>100	100	>100
	2	>100	100	>100

Table 3-9. Summary of Duplicate Test Analysis Using *Ceriodaphnia dubia*. Facilities 3 and 4 Were Unble to Conduct Duplicate Analyses.

Facility Code	Test	Three-Brood Survival			Three-Brood Reproduction		
		IC_{25} (%)	NOEC (%)	LOEC (%)	IC_{25} (%)	NOEC (%)	LOEC (%)
1	1	>100	100	>100	>100	100	>100
	2	>100	100	>100	>100	100	>100
2	1	>100	100	>100	>100	100	>100
	2	>100	100	>100	>100	100	>100
3	1	>100	100	>100	>100	100	>100
	2	NM	NM	NM	NM	NM	NM
4	1	>100	100	>100	>100	100	>100
	2	NM	NM	NM	NM	NM	NM
5	1	>100	100	>100	>100	100	>100
	2	>100	100	>100	>100	100	>100
6	1	>100	100	>100	>100	100	>100
	2	>100	100	>100	>100	100	>100

Table 3-10. Summary of Duplicate Test Analysis Using *Pimephales promelas*. Facilities 3, 4, and 5 Were Unable to Conduct Duplicate Analyses.

Facility Code	Test	7-D Survival			7- D Growth		
		IC_{25} (%)	NOEC (%)	LOEC (%)	IC_{25} (%)	NOEC (%)	LOEC (%)
1	1	>100	100	>100	>100	100	>100
	2	>100	100	>100	>100	100	>100
2	1	>100	100	>100	>100	100	100
	2	>100	100	>100	>100	100	>100
3	1	85.7	100	>100	67.1	50	100
	2	NM	NM	NM	NM	NM	NM
4	1	>100	100	>100	>100	100	>100
	2	NM	NM	NM	NM	NM	NM
5	1	NM	NM	NM	NM	NM	NM
	2	NM	NM	NM	NM	NM	NM
6	1	>100	100	>100	>100	100	>100
	2	>100	100	>100	>100	100	>100

Table 3-11. Summary of Duplicate Test Analysis Using *Selenastrum capricornutum*. Facilities 1, 2, and 4 were Unable to Conduct Duplicate Analyses.

Facility Code	Test	96-H Growth		
		IC_{25} (%)	NOEC (%)	LOEC (%)
1	1	NM	NM	NM
	2	NM	NM	NM
2	1	29.5	12.5	25
	2	NM	NM	NM
3	1	7.33	<6.25	6.25
	2	16.7	<6.25	6.25
4	1	8.2	<29.5	29.5
	2	NM	NM	NM
5	1	19.6-	12.5-	25-
	2	NM	NM	NM
6	1	>100	100	>100
	2	>100	100	>100

Table 3-12. Number (and Percentage) of WET Tests that Would Not Meet their Compliance Limit Based on Either Design Low Flow or Actual Flow Conditions at the Time of Testing, and Based on Either the No Observed Effect Concentration (NOEC) or the IC_{25} Value for Each Test Type.

Facility	*C. dubia*		*P. promelas*		*S. capricornutum*	
	NOEC	IC_{25}	NOEC	IC_{25}	NOEC	IC_{25}
1	1 (25)	1 (25)	0 (0)	1 (25)	2 (100)	2 (100)
2	3 (25)	2 (17)	6 (50)	4 (33)	3 (100)	3 (100)
3	1 (25)	0 (0)	3 (75)	2 (50)	3 (100)	2 (67)
4	0 (0)	0 (0)	0 (0)	1 (25)	3 (100)	3 (100)
5	0 (0)	0 (0)	0 (0)	0 (0)	3 (100)	3 (100)
6	0 (0)	0 (0)	1 (20)	1 (20)	0 (0)	0 (0)
Total	5 (14)	3 (8.5)	10 (30)	7 (21)	14 (74)	13 (68)

Table 3-13. Summary of WET Testing Results by Measurement Quality Objective. Bold Entries Indicate Non-compliance with MQO. Numbers Represent the Number of Times MQO was not Met (in bold) Followed by the Total Number of Values Generated by that Facility. NA = not applicable.

Measurement Quality Objectives		Facility 1	2	3	4	5	6
Control Survival	Pp[1] > 80%	0 / 4	0 / 12	0 / 4	0 / 4	0 / 2	0 / 5
	Cd[2] > 80%	0 / 4	**1** / 12	0 / 5	0 / 4	0 / 4	0 / 5
Mean Individual Fish Weight after 7-d	≥ 0.25 mg dry weight	0 / 4	0 / 12	0 / 4	0 / 4	0 / 2	0 / 5
C. dubia offspring per female	≥ 15 offspring per surviving female	0 / 4	0 / 12	0 / 5	0 / 4	0 / 4	0 / 5
	60% of females with ≥ 3 Broods	0 / 4	0 / 12	0 / 5	0 / 4	0 / 4	0 / 5
S. capricornutum mean cell density in control	≥ 1 x 10^6 cells/ml	0 / 2	0 / 12	0 / 3	0 / 4	0 / 3	0 / 5
Percent minimum significant difference (PMSD) in P. promelas growth	< 35%	0 / 4	0 / 12	0 / 4	**1** / 4	0 / 2	0 / 5
PMSD for C. dubia reproduction	< 37%	**1** / 4	**3** / 12	**1** / 5	**3** / 4	0 / 4	0 / 5
Maximum coefficient of variation (CV) between S. Capricornutum[3] control replicates	< 20%	**1** / 2	0 / 3	0 / 3	0 / 3	0 / 3	0 / 5
CV for monthly reference toxicant test endpoints	Cd Survival CV < 35%	Yes	NA	Yes	**No**	Yes	Yes
	Cd Reproduction CV < 35%	**No**	NA	Yes	**No**	Yes	Yes
	Pp Survival CV < 35%	Yes	NA	Yes	Yes	**No**	Yes
	Pp Growth CV < 35%	Yes	NA	Yes	Yes	Yes	Yes
	Sc Growth CV < 35%	NA	NA	NA	NA	NA	Yes
Split toxicity treatments: % Difference for endpoints	Cd Survival PD < 35%	Yes	Yes	Yes	Yes	Yes	Yes
	Cd Reproduction PD < 35%	Yes	**No**	Yes	Yes	Yes	No
	Pp Survival PD < 35%	Yes	Yes	Yes	Yes	Yes	Yes
	Pp Growth PD < 35%	**No**	**No**	Yes	Yes	Yes	**No**
	Sc Growth PD < 35%	**No**	**No**	**No**	**No**	**No**	Yes
Split testing of method blank between labs: % Difference of endpoints	Cd Survival PD < 35%	Yes	Yes	NA	Yes	Yes	Yes
	Cd Reproduction PD < 35%	Yes	Yes	NA	Yes	Yes	Yes
	Pp Survival PD < 35%	Yes	Yes	NA	Yes	Yes	Yes
	Pp Growth PD < 35%	Yes	Yes	NA	Yes	Yes	Yes
	Sc Growth PD < 35%	**No**	**No**	**No**	**No**	**No**	**No**
Associated reference toxicant test yields endpoint of the long-term average	Cd Survival ± 35%	**2** / 12	NA	**1** / 12	**1** / 10	0 / 4	**1** / 13
	Cd Reproduction ± 35%	**6** / 12	NA	0 / 12	**9** / 10	0 / 4	**1** / 13
	Pp Survival ± 35%	**1** / 12	NA	0 / 14	**2** / 11	**15** / 27	**2** / 13
	Pp Growth ± 35%	**1** / 12	NA	**1** / 14	**2** / 11	**7** / 27	0 / 13
	Sc Growth ± 35%	NA	NA	NA	NA	NA	**2** / 7

[1]Pp = *Pimephales promelas*
[2]Cd = *Ceriodaphnia dubia*
[3]Sc = Selenastrum capricornutum

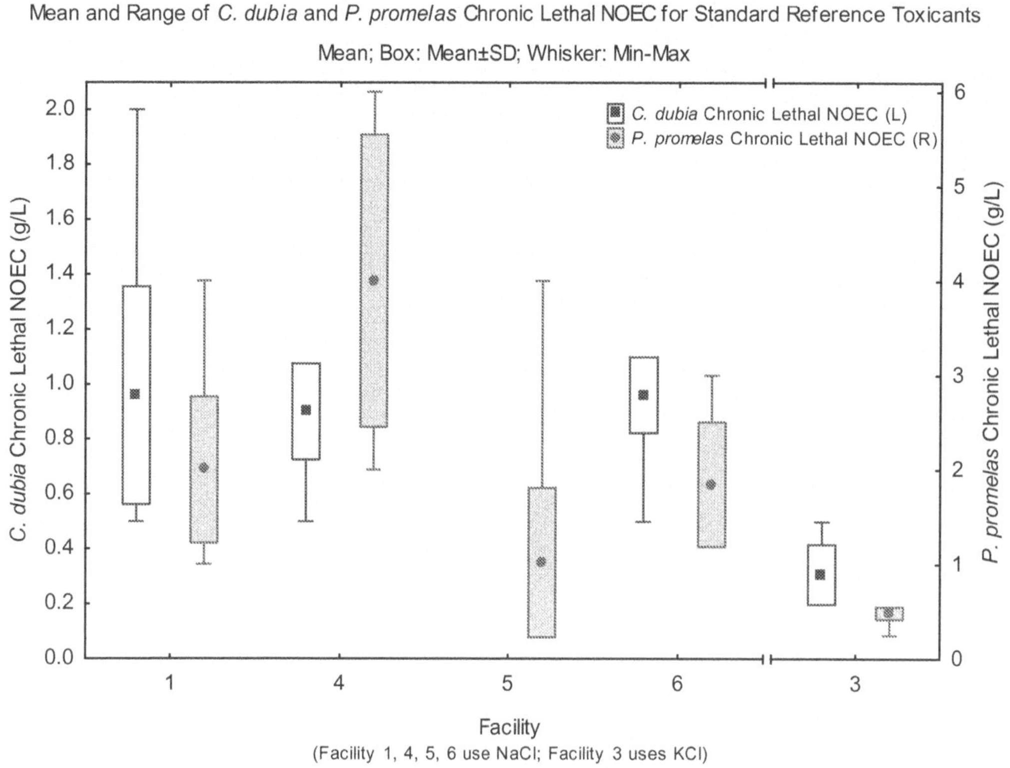

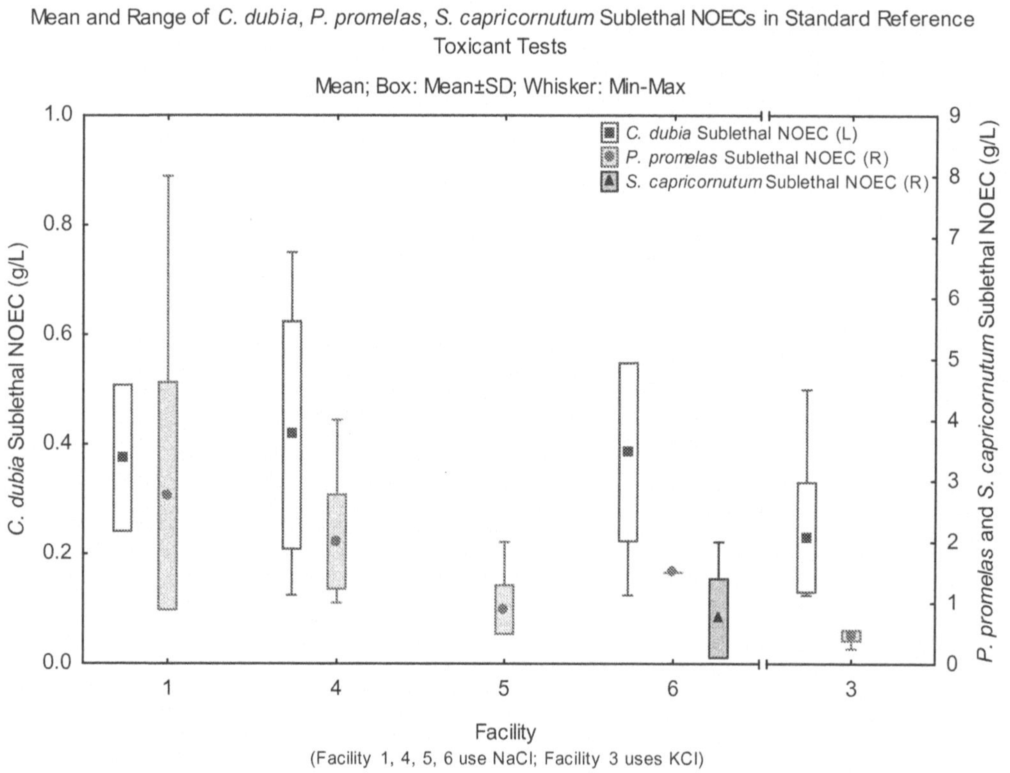

Figure 3-1. Summary of Reference Toxicant Test Results for Each Species Using NOEC Values.

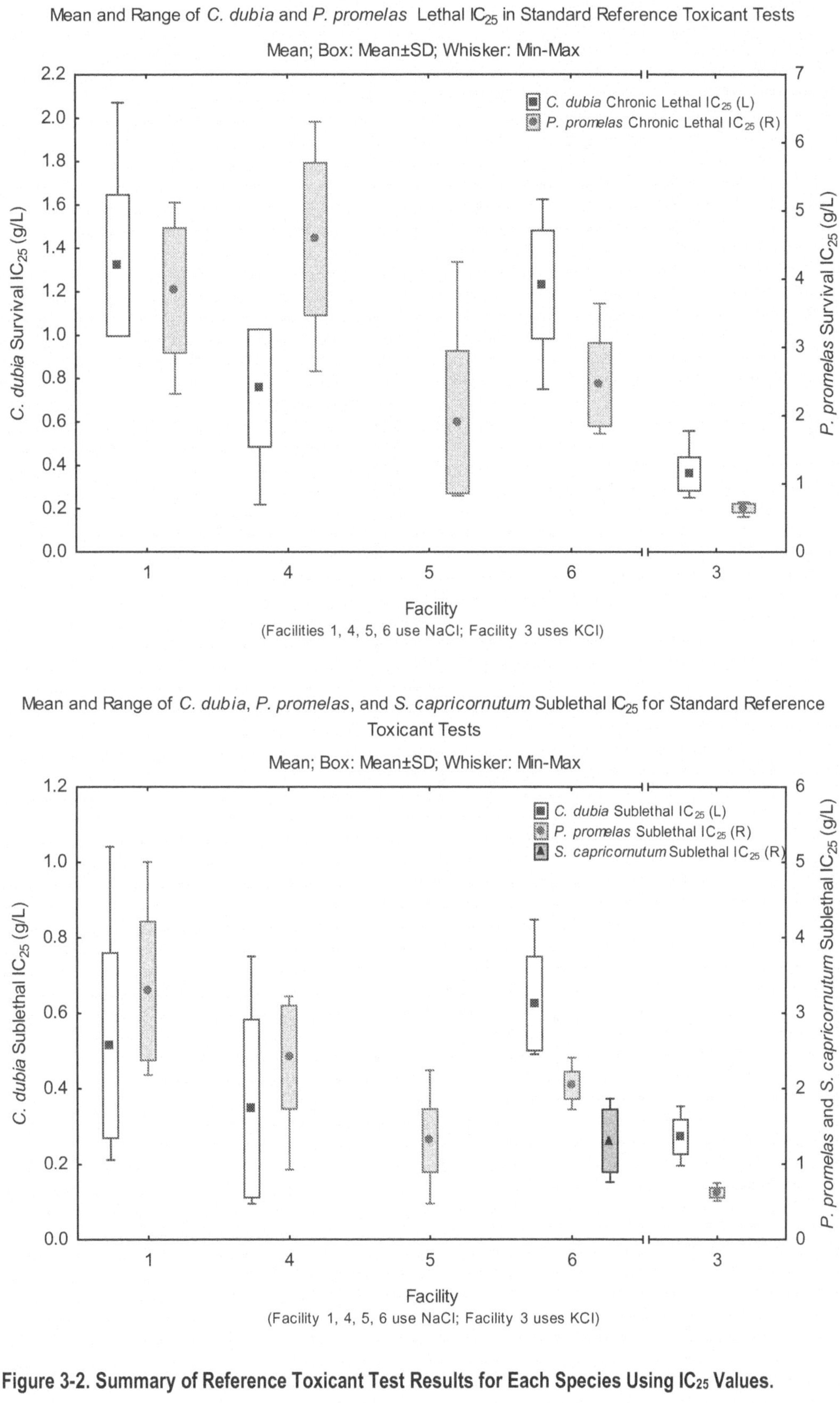

Figure 3-2. Summary of Reference Toxicant Test Results for Each Species Using IC_{25} Values.

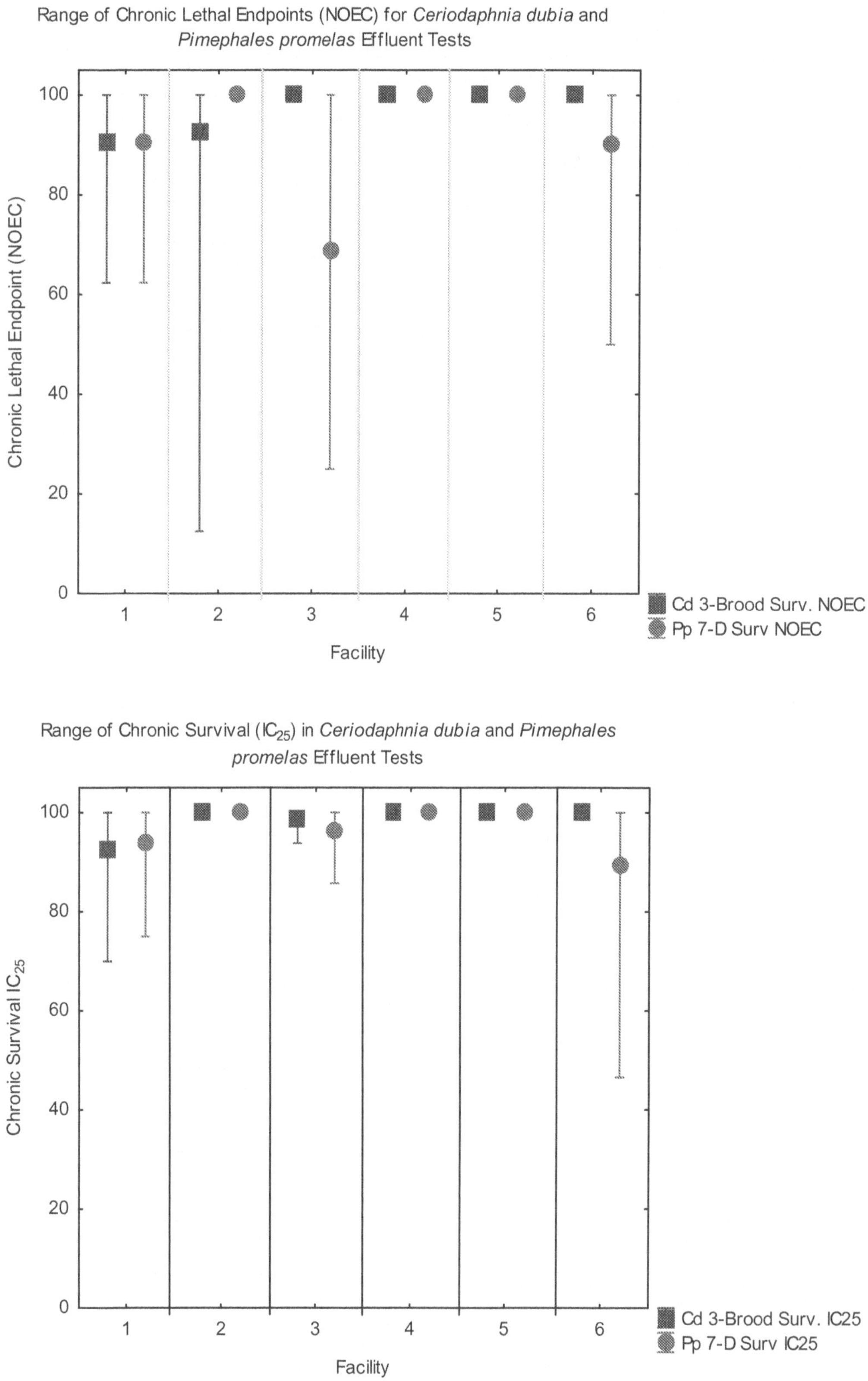

Figure 3-3. Mean and Range of Survival NOEC and IC_{25} Endpoint Values Observed for Each Facility Over a Year of Testing.

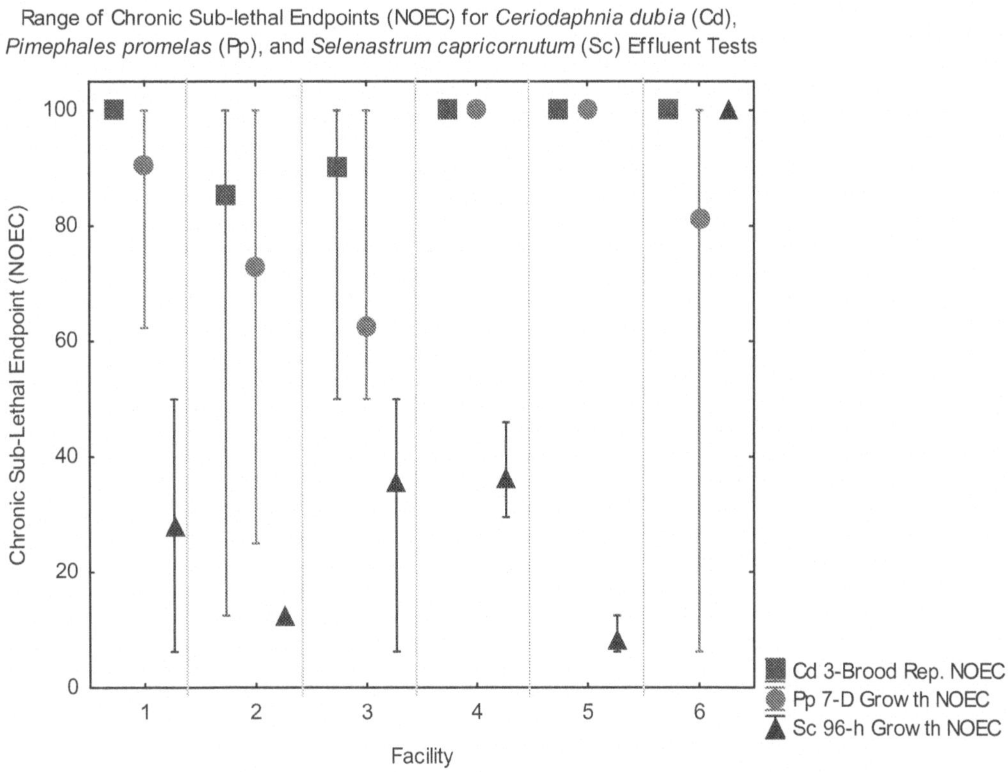

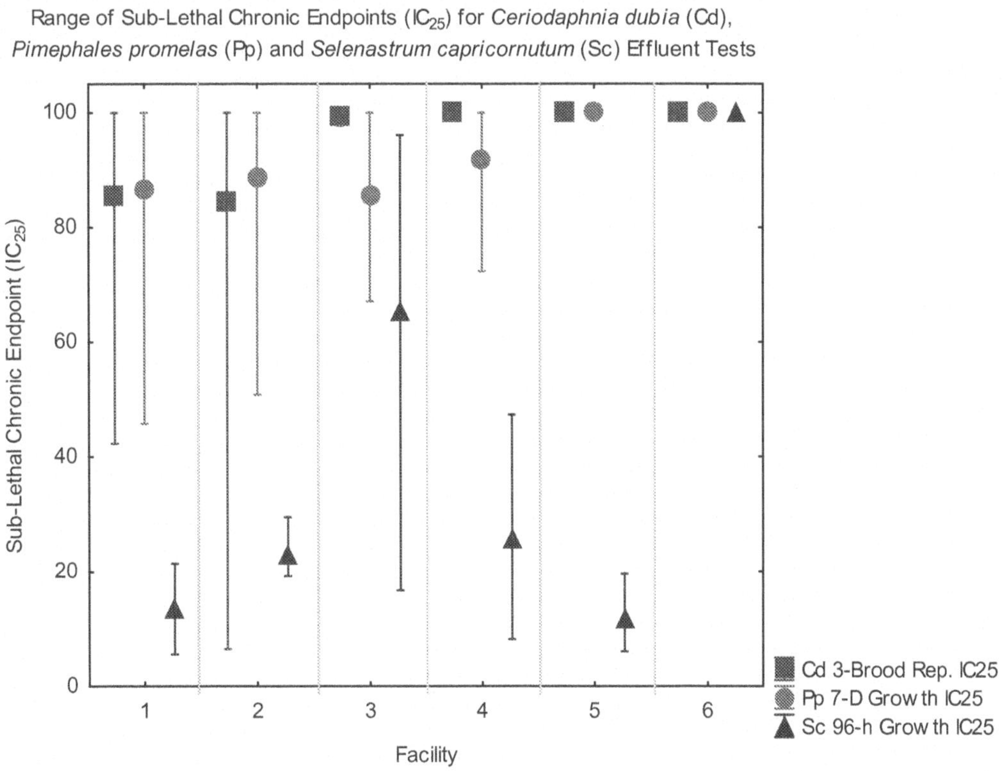

Figure 3-4. Mean and Range of Sub-lethal NOEC and IC_{25} Endpoint Values Observed for Each Facility Over a Year of Testing.

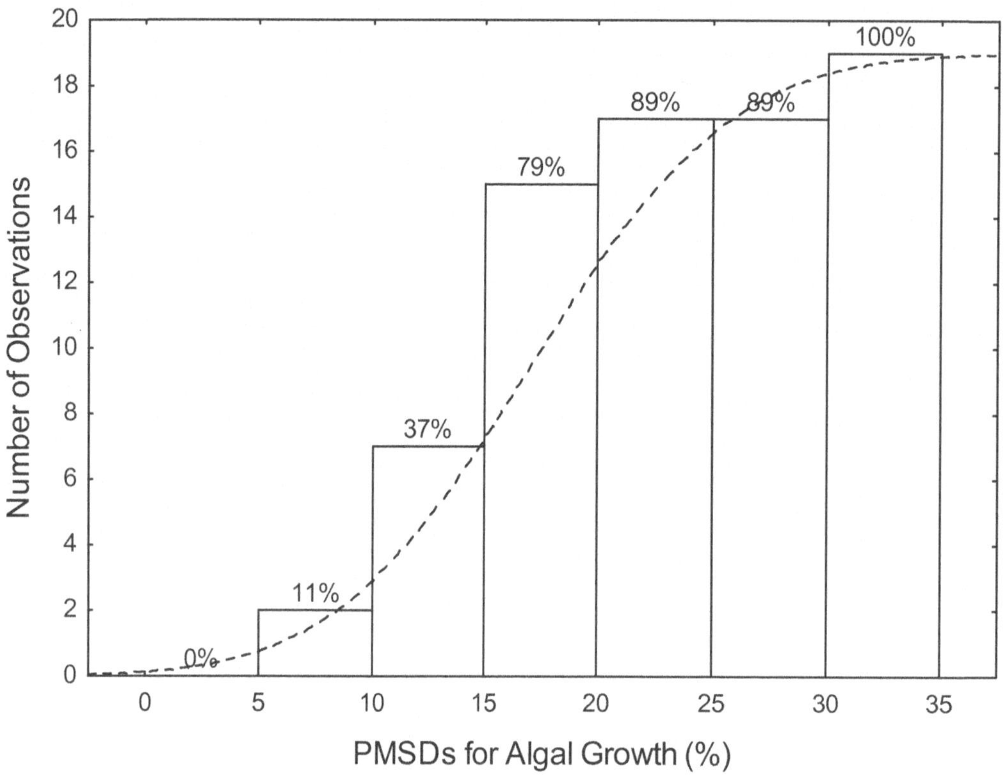

Figure 3-5. Cumulative Frequency Distribution of Percent Minimum Significant Differences (PMSDs) Calculated for Algal Growth for Each *Selenastrum capricornutum* Test Based on Control Replicate Data Observed. The Two Laboratories Had a PMSD < 30% for Algal Growth.

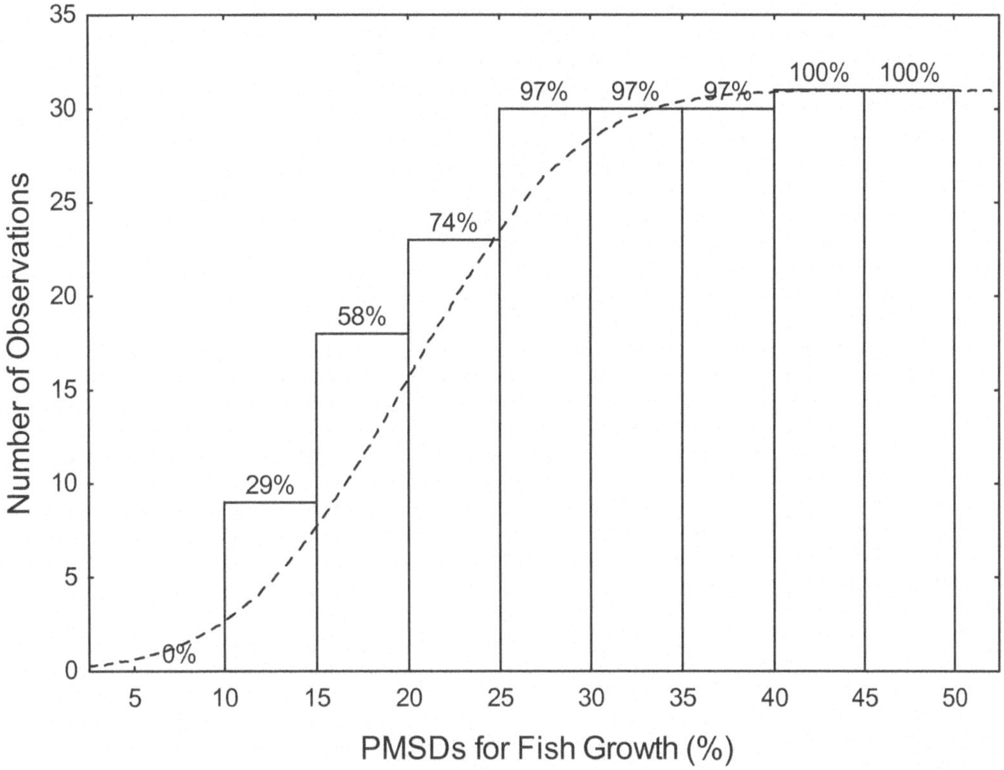

Figure 3-6 Cumulative Frequency Distribution of Percent Minimum Significant Differences (PMSDs) Calculated for Fish Growth for Each *Pimephales promelas* Test Based on Control Replicate Data Observed. Most Laboratories Had a PMSD < 30% for Fish Growth.

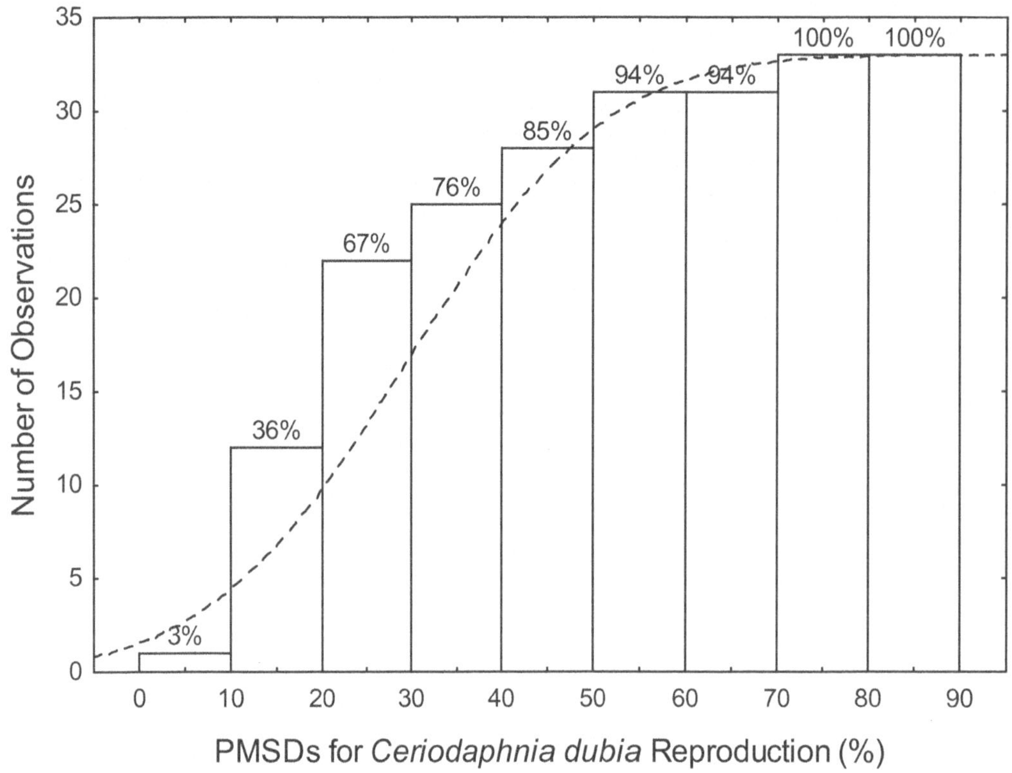

Figure 3-7. Cumulative Frequency Distribution of Percent Minimum Significant Differences (PMSDs) Calculated for Reproduction for Each *Ceriodaphnia dubia* Test Based on Control Replicate Data Observed. Most Laboratories Had a PMSD < 50% for *C. dubia* Reproduction.

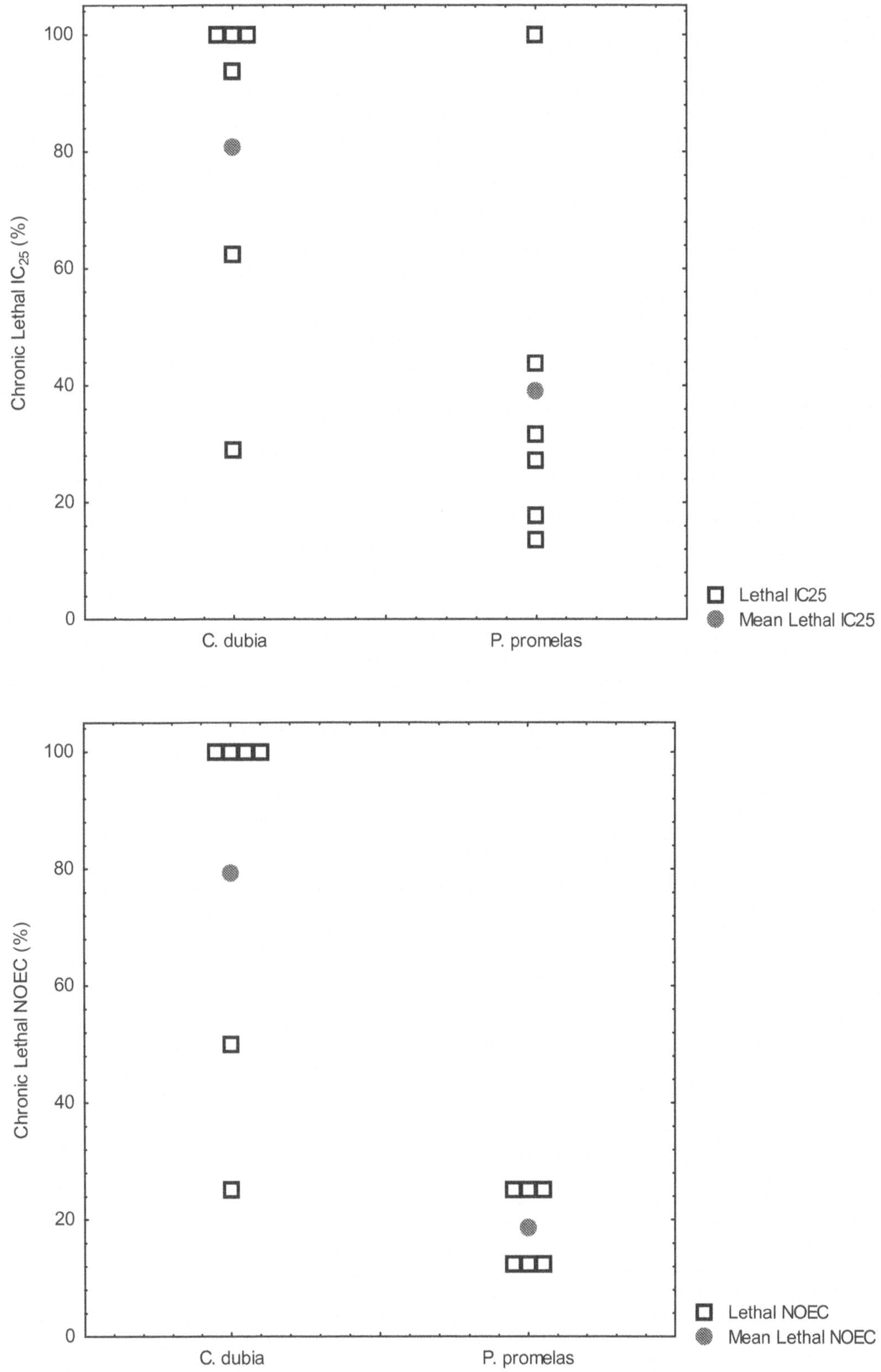

Figure 3-8. Survival NOEC and IC_{25} Values Reported by Each Facility, as Well as the Observed Mean Value, for *C. dubia* and *P. promelas* Tests Using Blind Sample Containing 16 or 128 ug/L Copper in the 100% Sample for the *C. dubia* and *P. promelas* Tests, Respectively. Each Laboratory Used Moderately Hard Reconstituted Water as Diluent (Hardness = 80-100 mg/L as $CaCO_3$)

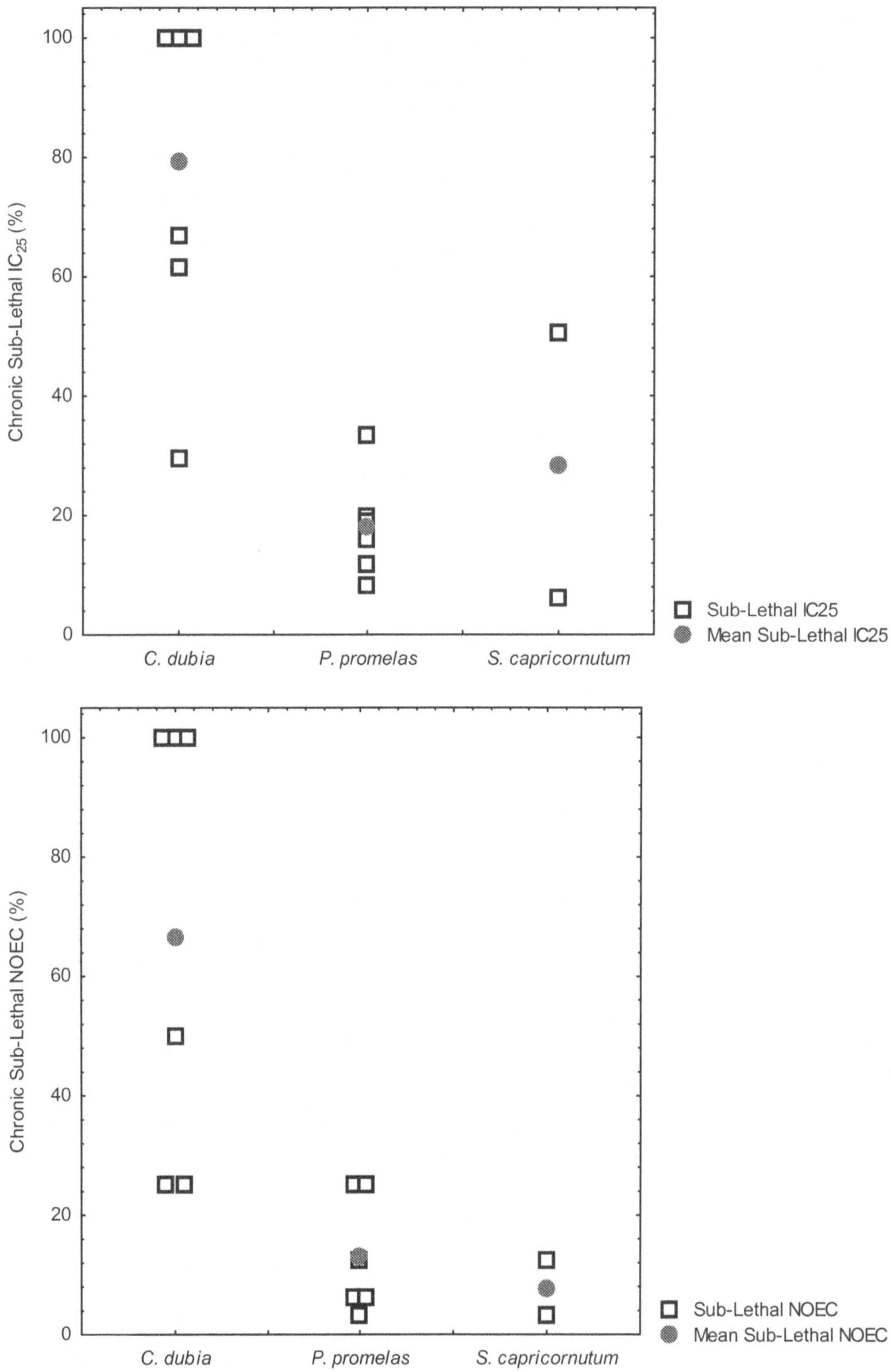

Figure 3-9. Sub-lethal NOEC and IC_{25} Values Reported for *C. dubia*, *P. promelas*, and *S. capricornutum* by Each Facility Using a Blind Sample Containing 16, 128, or 128 ug/L Copper in the 100% Sample for the *C. dubia*, *P. promelas*, and *S. capricornutum* Tests, Respectively. Each Laboratory Used Moderately Hard Reconstituted Water as Diluent (Hardness = 80-100 mg/L as $CaCO_3$).

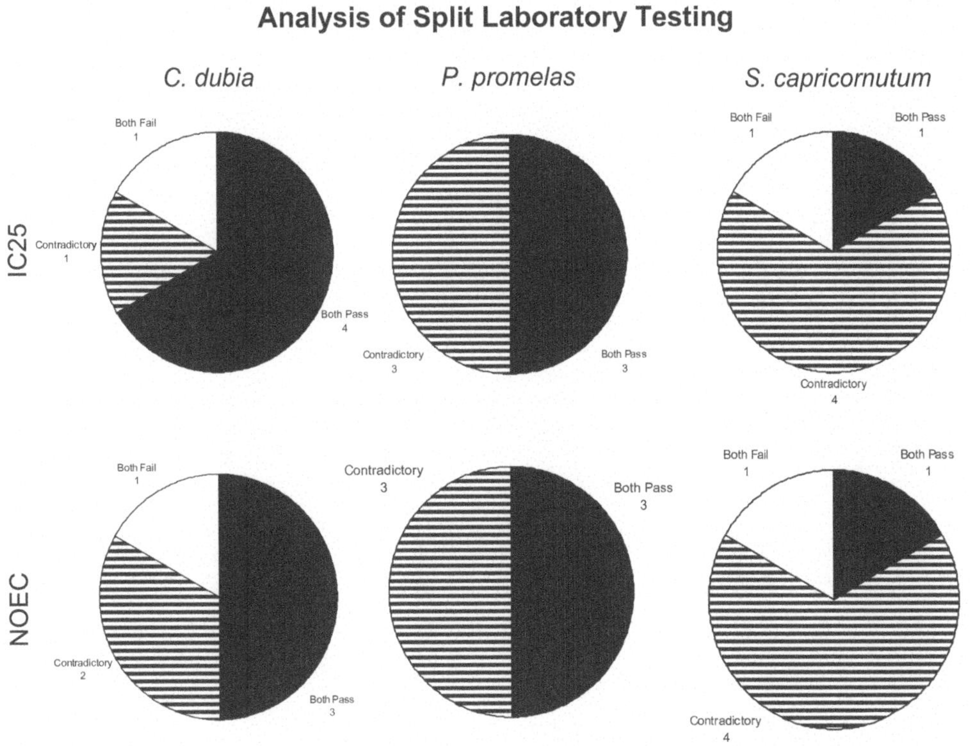

Figure 3-10. Comparison of split Testing Results in Terms of Test Passing or Failing Permit Limits, With Respect to IC_{25} and NOEC Values Reported by the Laboratories for Each Type of Test: *C. dubia, P. promelas and S. capricornutum.*

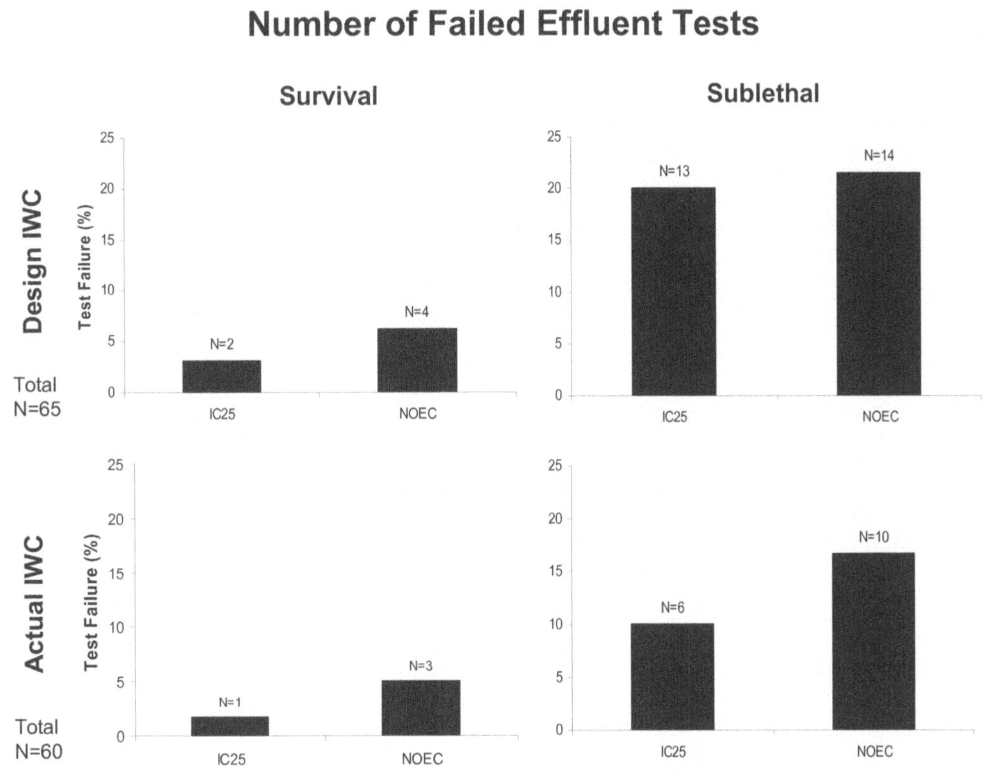

Figure 3-11. Comparison of Test "Failures" (i.e., Non-compliance With WET Limit) Based on Design and Actual Flow Conditions.

CHAPTER 4.0

BIOASSESSMENT RESULTS

As explained in the methods section of this report, three different bioassessments were conducted at each facility: periphyton, macroinvertebrates, and fish. In addition, two different sampling methods were evaluated for the macroinvertebrate bioassessment at two facilities in order to determine whether one or both methods might be needed in a more comprehensive study. The objectives of the bioassessment analyses in this project were: 1) to determine the performance of the different bioassessment methods (i.e., compliance with suggested MQO) and thereby determine what methods should be used in a more comprehensive study, and 2) determine the types of analytical methods that should be considered to identify ecologically significant changes in biota downstream of a discharge. To address these objectives, the project team analyzed the data for each type of bioassessment using two different and respected approaches: a) multivariate statistical analyses that examined the similarity of samples, in terms of species composition or metric values (if available), to determine method precision and differences between up and downstream samples at a facility, and b) multi-metric analyses to determine compliance with MQO as well as to examine and display differences between up and downstream samples. The following summarizes the results of these analyses for each assemblage.

4.1 Periphyton

4.1.1 Periphyton Bioassessment Performance

Precision between upstream replicate samples was within the MQOs proposed for many of the sites. Taxa richness, percent sensitive taxa, and percent tolerant taxa had Coefficients of Variation (CV) between 2 and 30% for most facilities. Facility 2 had substantially higher CVs for sensitive or tolerant taxa metrics than other facilities (CVs >79-87%). Taxa richness had the lowest CV for all facilities, with the highest CV being 26.5% (Facility 5).

Over 90% of the algal abundance for each site was attributed to diatoms for all facilities. Algae, including green, bluegreen, and red algae were also present at most sites, but in relatively small proportions. No major changes in functional groups were observed between upstream and downstream sites (see Appendix I, available by request, for periphyton data).

4.1.2 Multivariate Analysis

An 'A' statistical test (McCune and Grace, 2002) was used to determine whether species composition at upstream and downstream sites were significantly different. In this test, a *p*-value and an 'A' value are provided to determine the probability of difference among groups, and evaluate within group variation in the MRPP. A = 0 when heterogeneity within groups equals the expectation by chance, and A $\simeq$ 1 indicates that samples are very similar to each other. Ideally, the 'A' value should be close to 1.0 if periphyton communities at upstream and downstream sites are very similar. However, natural variability, as well as various sources of method error, typically result in lower 'A' values even for field replicate samples that should be nearly identical (McCune and Grace, 2002). 'A' values $\geq$ 0.50 generally indicate fairly high similarity between typical field samples.

Algal species composition in upstream sites differed from that of downstream sites in most of the study facilities (Table 4-1). Algal species composition at Facilities 1, 2, 3, 4, and 6 were significantly different between upstream and downstream samples ($p<0.05$). The A values for Facilities 2 through 4 were relatively higher, indicating fairly small within group variability at these sites; i.e., replicate samples within upstream and within downstream samples at these facilities were not very different. On the other hand, Facilities 1 and 6 had low 'A' values indicating that algal samples collected upstream and downstream of these facilities had relatively large variability in species composition. The significant p value observed for these sites, however, shows that this analysis could still detect a change in periphyton species composition between upstream and downstream samples at these facilities. Facility 5 had a similar 'A' value as Facility 1 but a non-significant p value, indicating that any supposed difference in algal composition between upstream and downstream samples at this site was within the range of differences observed among upstream samples alone.

4.1.3 Metric Analysis

Periphyton species richness was significantly lower downstream than upstream at Facilities 2 and 3 (Figure 4-1) and the percentage of tolerant diatom taxa increased downstream of Facilities 2, 3, and 4 (Figure 4-2). The percentage of sensitive diatom species was also lower downstream of Facilities 4 and 6 (Figure 4-2). Together, these data suggest that the species composition differences observed at Facilities 2, 3, and 4 between up and downstream, based on the multivariate similarity approach above (Table 4-1), appear to be related to what are typically considered non-desirable changes in the periphyton assemblage. Facility 5, which did not exhibit significant taxonomic composition differences up and downstream, did show a significant increase in algal cell density downstream (Figure 4-1). Based on the metric and multivariate results observed, Facilities 2, 3, 4, and 6 appeared to exhibit some decrease in periphyton condition downstream while Facilities 1 and 5 had relatively little detectable change in the periphyton assemblage up and downstream.

4.2 Benthic Macroinvertebrates

4.2.1 Laboratory Performance

Prior to analysis, the performance of macroinvertebrate subsampling and taxonomic identification procedures were documented for compliance with MQOs. Three MQO were applied (see Methods section): sorting bias (PSE), enumeration (PDE), and taxonomic accuracy (PTD). For the project overall, the sorting and subsampling MQO (for bias) was met with a PSE of 95.4% (Table 4-2; see Appendix J, available by request, for macroinvertebrate data). Laboratories A and B performed sorting and subsampling. Laboratory A did not have an external sort re-check for this project, but has documented the attainment of a PSE of 96.9% for another recently completed project; this demonstrates the capacity of the laboratory to appropriately apply the sorting and subsampling procedures. Laboratory B, which performed sample sorting and taxonomy for Facility 6, initially failed the sorting MQO, having a PSE of 62.4%. As the corrective action, Laboratory B re-sorted all samples, ultimately resulting in compliance with the MQO (PSE = 95.4%).

The precision of taxonomic identifications for the project overall was met with PDE = 1.8% and PTD = 10.5%. Individually, taxonomy was performed by Laboratories B and C. The former initially did not meet the precision objective for either the enumeration or the taxonomy, which, in large part, was due to the incomplete sample being transferred to the independent taxonomist for reprocessing. The corrective action was to make the two sets of results comparable (strictly for the QC analysis) by excluding from consideration those 75 specimens, mostly slide-mounted midges (Insecta: Diptera: Chironomidae). This resulted in an improvement of taxonomic precision to 1.8% PDE (from 26.9) and 10.5% PTD (from 54). Laboratory C taxonomic performance is documented for a previous, recently completed project (Gallardo et al. 2006), specifically for taxonomists performing identifications in this project. In that comparison, initial comparisons showed PTD as exceeding the MQO at approximately 29%, primarily due to hierarchical differences. Adjustment for samples having an abundance of early instar, damaged mayflies (Insecta: Ephemeroptera) resulted in a PTD of 4.9%.

4.2.2 Field Sampling Performance (Benthic Macroinvertebrates)

For all facilities, precision of field sampling was calculated by using three upstream samples (Facilities 1-5) or two (Facility 6). The standard deviation, root mean squared error, coefficient of variability, and the 90% detectable difference were calculated for each facility bioassessment (Table 4-3). Precision for benthic sampling met the MQO at all facilities, including Facility 6 (CV=7.2%) for which there were two "reference" sites independent of the receiving water. For the others, which all relied on three upstream samples, precision was even tighter, with CV ranging from 0.9-2.9%. The differences that could be reliably distinguished with 90% confidence (DD_{90}) for all facilities ranged from ±0.6-1.4 index score. All facilities exhibited at least three macroinvertebrate metrics with CV <10%, again indicating high precision. Among the metrics used in this Pilot, percent Ephemeroptera and percent scrapers were consistently the least precise for most facilities, with CVs ranging from 20% up to 162%. A full study would need to use calibrated metrics for the ecoregion in which a facility is located.

4.2.3 Multivariate Analysis

Macroinvertebrate taxonomic composition was significantly different between upstream and downstream locations at Facilities 2, 4, 5, and 6 (Figure 4-3). The 'A' values for Facilities 2, 4 and 5 were relatively high indicating fairly low within group variability; i.e., replicate samples within upstream and downstream locations were similar (Table 4-4). Facility 5 had an 'A' value of 0.5 indicating very low variability in species composition within upstream or downstream locations. Although Facility 6 had a low 'A' value (high within group variability), this statistical approach could still detect a difference between the downstream and reference locations. Facility 1 had a very low 'A' value (≈ 0) indicating high within group variability in species composition that would tend to mask actual differences between upstream and downstream locations at that Facility.

On a metric basis (examining all of the best candidate metrics simultaneously in this analysis), the macroinvertebrate characteristics in upstream locations were different than that in downstream stations in three of the six facilities. Similar to results obtained based on species composition, metrics values for Facilities 2, 4, and 6 were significantly different between upstream and downstream locations (Table 4-5).

4.2.4 Multi-Metric Analyses

The normalized composite indexes were compared in an upstream-downstream manner for each facility (Figures 4-3 to 4-8). Mean upstream index scores are taken to be representative of the biota of those stream reaches, for the particular sampling technique used (either multihabitat dipnet or artificial substrate Hester-Dendy). As noted, the "upstream" for Facility 6 is actually represented by two reference sites that are located on streams different than the receiving waters. Upstream-downstream differences are strongly evident for Facilities 2 and 4 (Figures 4-4 and 4-6), similar to results observed using multivariate analysis and not as consistent or distinct for the other facilities. Composite indexes for each of Facilities 2 and 4 had DD_{90} of ±1.0 and 0.6, respectively (Table 4-3). For the former, no downstream site is closer than approximately 20 points below the upstream mean. For Facility 4, no site was closer than 25 points below the upstream mean. For Facility 2, there were strong upstream-downstream differences for seven out of the 11 metrics (Table 4-6), which are principally taxa richness-type metrics and tolerance metrics (total taxa, number of EPT taxa, percent of EPT individuals, percent of Ephemeroptera individuals, number of intolerant taxa, percent of tolerant individuals, and percent individuals of most dominant taxon). Similar metrics showed a very strong upstream-downstream difference for Facility 4 (Table 4-6). At this particular facility, two metrics strongly influenced by taxa with the habit designated as clinger (number of clinger taxa, and percent clinger individuals) also showed a very strong difference, and could potentially be reflecting differences in habitat. Although all other facilities have very tight precision estimates (Table 4-3), there was only very slight suggestion of downstream differences if any.

4.2.5 Comparison of Macroinvertebrate Sampling Methods: Multivariate Analysis

Two macroinvertebrate sampling methods (Dipnet and Hester-Dendy) were compared in terms of field sampling precision. At Facility 6, the two reference site Dipnet samples exhibited a difference in species composition corresponding to a dissimilarity index of 0.55 while the two reference site Hester-Dendy samples had a mean dissimilarity index of 0.65, indicating that precision was somewhat better using the Dipnet sampling method at these sites (Figure 4-9). The mean difference in macroinvertebrate composition between the down stream sites and the reference sites (0.69 ± 0.04 for Dipnet, and 0.77 ± 0.09 for Hester-Dendy) was greater than the difference between reference sites using both collection methods. This indicates that both sampling methods could detect some change in macroinvertebrate composition downstream of Facility 6 as compared to the reference sites.

Three pairs of samples were collected using both sampling methods upstream and downstream of Facility 5. The mean difference among upstream sites was not significantly different between methods (Figure 4-9). However, precision appeared to be better using the Dipnet method (i.e., lower dissimilarity: 0.49 and 0.63 using Dipnet and Hester-Dendy respectively). The difference in macroinvertebrate composition between samples collected up and downstream of this Facility was similar between the two methods (0.55 ± 0.10 and 0.60 ± 0.35 for the Dipnet and Hester-Dendy method, respectively). However, in general, there was greater variability in species composition among up and downstream sites based on Hester-Dendy sampling than Dipnet sampling (i.e., larger error bars for the Hester Dendy dissimilarity).

The above results, though limited, suggest that the Hester-Dendy sampling method yielded somewhat less precision in terms of species composition between field replicates and more variability in terms of up and downstream comparisons than the Dipnet method.

4.2.6 Comparison of Macroinvertebrate Sampling Methods: Multimetric Analysis

Precision of Composite Indexes. Even with such small sample sizes, both techniques demonstrated low Coefficients of Variation (CV) indicating good precision and high power to detect differences (Figure 4-10). However, within this sample set, for each facility the precision using the dipnet method was better than that using Hester Dendy samplers. For Facility 5 that difference was substantial, but may reflect a level of precision that is not always necessary (from a cost-benefit perspective). The difference in precision of the two methods at Facility 6 (Figure 4-10), while evident, is not large and is potentially confounded by the fact that the two reference samples are from off-stream sites.

Similarity of Metric Values. The 11 individual metrics varied in how similar their precision was between the two sampling methods, with average relative percent difference (RPD) ranging between 34.5% and 142.7%. Metrics with the greatest similarity were percent dominant taxon (Dom01Pct), number of taxa (Figure 4-11), and percent filterers; those with the least similarity are percent Ephemeroptera, percent EPT, and percent scrapers (Figure 4-12). For seven out of 11 samples, number of taxa differed by 10 or less, and suggests that bias may be a minimal factor in how the techniques perform in these kinds of stream systems. Four metrics (number of clinger taxa, percent of clinger individuals, number of taxa, and percent dominant taxon had r^2 values $\geq$ 0.79, further suggesting that the latter two metrics are only slightly affected by gear differences. This also suggests that there are likely natural habitat features in these two streams that the artificial substrates closely mimic, such as cobble or other hard substrate.

Differential Responses to Effluent. Examining composite index scores downstream, specifically, the magnitude of change from the upstream mean (Figure 4-13), no distinct and consistent differences were apparent in how the methods characterize the sampling locations. At Facility 5, the dipnet assessments were consistent in being slightly lower than the mean. The Hester Dendy samples indicated the same assessment for the first two locations, but then is opposite for reach DS3. The situation is reversed for Facility 6, where the dipnet samples are higher than the upstream mean for all three downstream samples. While there may not be a sufficient number of samples or sites to be definitive, it appears that either approach could be used, with relatively equivalent confidence, to evaluate instream biological responses to discharges.

Other Considerations. The CV of the dipnet composite indexes, for both facilities, was lower than those produced by the artificial substrate samples indicating better precision using the dipnet method. Although this is a desirable data quality characteristic of biological assessment protocols, differences in the sampling and sample processing costs should also be considered. Just for laboratory processing time (sorting and subsampling), the difference in person-hours for the 200 organism subsamples to be attained was a total of 10 hours, so, on average, approximately two hours <u>more</u> per sample was required to sort a dipnet sample as opposed to a Hester-Dendy sample. This is largely due to the fact that samples coming into the laboratory for that phase of processing are much more difficult and slow to work with if there is an over-abundance of leaf litter and other detritus. Hester-Dendy samples, by the nature of how they work, almost always have much less detritus to be sorted. However, any colonization-based sampling method like Hester-Dendy samplers require two trips to the field for each site: one to set the samplers and one to retrieve them several weeks later. Aside from the added travel and field labor costs, artificial samplers are prone to vandalism or loss through sediment burial and flooding. In fact, one of the samplers in this study was lost through burial at Facility 5. Thus, depending on the nature of the stream habitat and hydrology, and the distance to the site, dipnet sampling could be less costly and ensure higher data completeness than Hester Dendy samplers.

The foregoing indicates the importance of the purpose of the sampling; that is, what the samples are intended to represent. The Hester-Dendy, in effect, supplies habitat and offers a consistent substrate for colonization (Resh and Rosenberg 1982). Thus benthic samples taken this way characterize the colonization potential of a waterbody. The dipnet approach samples the habitat that is available and captures the organisms that are there, whatever the habitat may be, even if the habitat is poor. The slightly better performance of dipnet sampling, as well as their ability to obtain truly representative taxonomic samples at a given site, may make the dipnet method preferable in studies of this kind.

4.3 Fish

4.3.1 Fish Bioassessment Performance

Precision of fish bioassessments based on upstream replicate samples was greater than desired by in large (see Appendix K, available by request, for fish data). The relative percent difference (RPD) of index values in upstream samples was often >50% at the six facilities, indicating that we were unlikely to be able to statistically detect less than a 40-50% change in fish index value between sites at a given facility. RPDs based on individual fish metric values were similar (RPD = 30-55%).

4.3.2 Multivariate Analysis

Due to the small sample size, differences in fish species composition between upstream and downstream locations could only be calculated at Facilities 3, 4, and 5. The 'A' values reported here are questionable because only two samples from the reference or upstream locations were collected for fish. While collection of two fish samples per reach is not unusual, given that this sampling is typically labor intensive and requires large reaches, fewer replicates limits the utility of a multivariate analysis approach.

Given the above caveat, the multivariate approach could not detect a difference in the fish species present downstream as compared to upstream at Facilities 3, 4, and 5 (Table 4-7). The difference in fish species composition between the upstream locations at Facilities 3 and 5 (0.421 and 0.340, respectively) was nearly the same as the mean difference between upstream and downstream locations at these sites (0.414 ± 0.087 for Facility 3 and 0.345 ± 0.075 for Facility 5), suggesting fairly poor discrimination ability of these data. However, the mean difference for downstream and upstream locations at Facility 4 was slightly larger than the upstream difference (0.294 ± 0.085 and 0.249, respectively), suggesting different fish communities upstream and downstream of this facility.

4.3.3 Index of Biotic Integrity Analysis

Eleven metrics commonly used in fish indices of biotic integrity (IBI) (Barbour et al., 1999) were calculated for each fish sample and each location. They were used to allow development of precision estimates and comparison of l potential downstream changes in biological condition relative to each facility. For each site, all 11 metrics scores were averaged to produce a site index score and corresponding 95% Confidence Intervals.

Only one facility location (Facility 6) showed significant differences between up and downstream index scores (Figure 4-14). This result was consistent with the multivariate results above. Downstream sites 1 and 2 had significantly lower site index values than one of the reference sites. This was the only facility that did not have a viable upstream flow and relied instead on two reference streams. At the remaining five facilities, there was no significant difference in the fish index scores between upstream and downstream sites (Figures 4-15 – 4-17), in contrast to the multivariate results, which indicated significant changes at Facilities 1 and 2.

Analysis of individual fish metric scores was also conducted for each facility to determine whether there were differences that may have been masked by the index calculation. Metric results are summarized in Appendix K. The following section summarizes fish results by facility.

Facility 1 Percent of the population as omnivores, top carnivores, creek chubs, and percent dominant taxa tended to be higher upstream of the outfall, suggesting some impairment upstream. The percent of the population as invertivores, total number of tolerant taxa, and total number of individuals captured appeared to be higher in sites downstream of the outfall. All other metrics show no apparent trends.

Facility 2 Percent omnivores appeared to be higher upstream of the outfall and percent insectivores appeared to be lower. All other metrics showed no apparent differences between up and downstream sites. Creek chubs were only found in the two sites directly down stream of the outfall.

Facility 3 The percent of the population as omnivores appeared to increase slightly downstream of the outfall, whereas the percent of the population as creek chub appeared to slightly decrease downstream of the outfall. The remaining metrics showed no obvious pattern.

Facility 4 There were no apparent differences between upstream and down stream sites with regards to the Shannon diversity index, percent omnivores, percent top carnivores, number of sucker species, percent creek chubs, number of sunfish species, percent dominant taxa, and total taxa. Percent insectivore and total number of individuals captured tended to increase downstream of the outfall.

Facility 5 There were no distinguishable differences between upstream and downstream sites with regard to any of the fish metric scores.

Facility 6 There were no notable differences between references sites and sites down stream of the outfall in terms of percent top carnivores, number of centrarchid species, and total number of individuals captured. Values for Shannon diversity index, percent omnivores, percent invertivores, total tolerant taxa, and total taxa tended to be higher in the references sites and increased with distance downstream of the outfall. Percent of population as creek chub was highest just below the outfall and lowest in the reference sites and in sites further from the outfall. Percent dominant taxa showed a similar trend.

Table 4-1. Comparison of Species Composition Similarity for Periphyton in Samples Collected Upstream Sites and Downstream of Each Facility. Bold Values Indicate Significant (P < 0.05) Differences Between Up and Downstream Periphyton Assemblages.

Facility	Mean Sorenson distance Upstream	Mean Sorenson distance Downstream	A Statistic	P Value
1	0.269	0.59	0.052	**0.045**
2	0.433	0.439	0.207	**0.024**
3	0.361	0.416	0.263	**0.022**
4	0.428	0.376	0.174	**0.025**
5	0.406	0.442	0.053	0.113
6	0.416	0.414	0.07	**0.045**

Table 4-2. Performance Characteristics for Laboratory Aspects of Benthic Macroinvertebrate Sample Processing. "Primary" and "Secondary" Indicates Lab Performance Pre-Corrective Action and Post-Corrective Action, Respectively.

Laboratory	Sorting (PSE)		Taxonomy			
			Enumeration (PDE)		*Identification (PTD)*	
	primary	secondary	primary	secondary	primary	secondary
A	96.9[1]	na	na	na	na	na
B	62.4	95.4	26.9	1.8	54	10.5
C[2]	na	na	0	na	21.8	4.9

[1]PSE attained in another project, presented to demonstrate laboratory capabilities

[2] no re-identifications performed for Lab C in this project; values are taken from recent but separate project involving Lab C taxonomist (see Gallardo et al. 2006)

Table 4-3. Precision Statistics for Benthic Macroinvertebrate Field Sampling (Dipnet Multihabitat) Using Upstream (or Reference) Samples for Calculations. Numbers are Based on the Composite Index for Each Facility.

Facility	n	$\bar{x}$	SD	RMSE	CV	DD_{90}
1	3	57.9	1.8	1.3	2.3	1.3
2	3	55.6	1.0	1.0	1.8	1.0
3	3	45.6	1.7	1.3	2.9	1.3
4	3	70.3	0.4	0.7	0.9	0.6
5	3	65.4	2.1	1.5	2.2	1.4
6	2	37.8	7.5	2.7	7.2	4.5

Table 4-4. Comparison of Species Composition Similarity for Macroinvertebrates in Samples Collected Upstream and Downstream of Each Facility. Facilities That Exhibited Significant Differences Between Up and Downstream Samples are Bolded (P< 0.05)

	Mean Sorenson distance			
Facility	Upstream	Downstream	A Statistic	P Value
1	0.329	0.463	-0.008	0.555
2	0.591	0.303	0.33	**0.023**
3	0.482	0.532	0.078	0.168
4	0.34	0.166	0.5	**0.022**
5	0.394	0.407	0.15	**0.035**
6	0.552	0.693	0.061	**0**

Table 4-5. Test of Significant Differences in Macroinvertebrate Metrics between Upstream and Downstream Stations. Bold Values Indicate Significant (P < 0.05) Differences Between Up and Downstream Macroinvertebrate Metrics.

	Euclidean Distance			
Facility	Upstream	Downstream	A Statistic	P Value
1	29.04	40.6	-0.047	0.727
2	39.37	27.48	0.458	**0.022**
3	36.42	42.1	0.058	0.311
4	17.85	16.18	0.617	**0.022**
5	26.45	43.9	0.161	0.121
6	28.17	44.11	0.116	**0**

Table 4-6. Upstream and Downstream Metric (Dipnet Only) Statistics from Facilities 2 and 4. "Direction" Means Expected Direction of Change in the Presence of Stressors.

			F2				F4			
			Upstream		Downstream		Upstream		Downstream	
Metrics	Direction	N	Mean	STD	Mean	STD	Mean	STD	Mean	STD
Total Taxa	▼	3	26	1.7	10	3.1	38	1.0	16	6.7
EPT Taxa	▼	3	4	0.6	1	1.0	10	0.6	3	1.2
EPT Percent	▼	3	34.2	8.5	0.6	0.7	46.8	6.1	31.9	5.6
Pct Ephemeroptera	▼	3	25.1	14.2	0.2	0.3	3.4	2.7	0.0	0.0
Intolerant Taxa	▼	3	3	1.5	0	0.6	9	1.7	2	1.0
Tolerant Percent	▲	3	17.8	7.4	76.9	16.1	4.1	1.9	1.9	1.8
Dom01Pct	▲	3	30.9	8.4	47.1	9.4	23.6	6.7	53.8	3.1
Filterers Percent	▲	3	8.6	8.2	1.8	1.0	49	6.1	31.9	6.0
Scraper Percent	▲	3	0.3	0.3	0.0	0.0	11.6	2.8	2.0	0.6
Clinger Taxa	▼	3	4	0.6	3	1.0	20	1.7	6	1.5
Clinger Percent	▼	3	30.4	19.6	34.5	8.5	75.5	4.9	37.3	5.1
Composite Index	▼	3	55.6	1.0	29.0	4.6	70.3	0.4	41.1	3.3

Table 4-7. Comparison of Fish Species Composition Similarity in Samples Collected Upstream and Downstream of Each Facility. Facilities That Exhibited Significant Differences between Up and Downstream Samples are Bolded (P< 0.05). Facilities With No A or P Values Indicates Not Enough Samples to Analyze Using this Approach.

	Mean Sorenson distance			
Facility	Upstream	Downstream	A Statistic	P Value
1	0.165	0.375	0.205	**0.000**
2	0.709	0.522	0.082	**0.042**
3	0.421	0.311	-	-
4	0.249	0.269	-	-
5	0.34	0.397	-	-
6	0.607	0.633	0.168	**0.000**

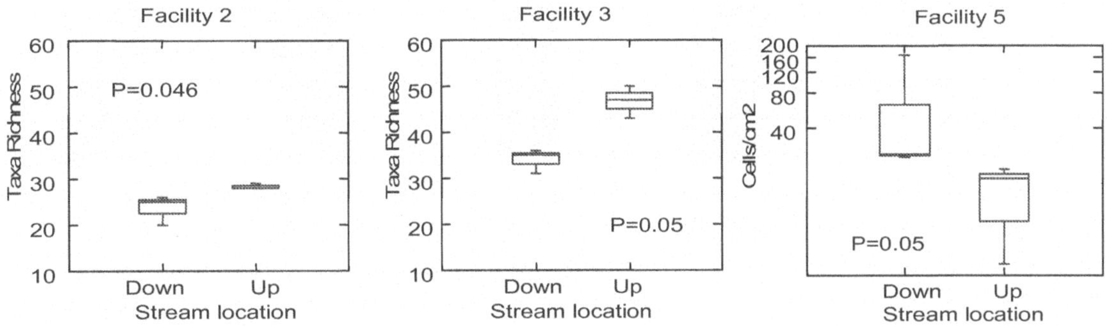

Figure 4-1. Comparison of Taxa Richness and Algal Cell Density Observed in Periphyton Samples Collected Upstream and Downstream of Select Facilities. Nonmetric Mann-Whitney U Test Was Used to Test Differences of Two Groups.

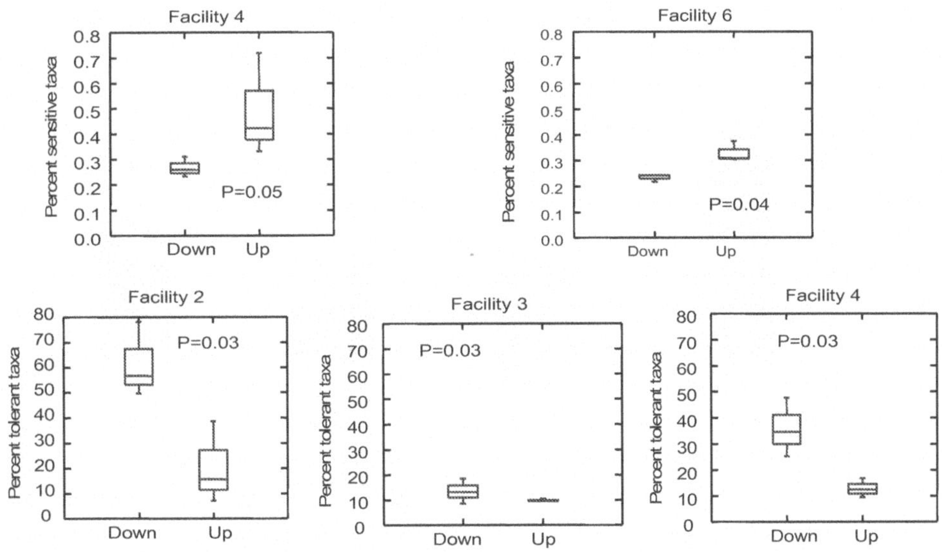

Figure 4-2. Comparison of Percent Tolerant or Intolerant Periphyton Species Observed in Periphyton Samples Collected Upstream and Downstream of Select Facilities. Percent Tolerant and Sensitive Algal Taxa Were Determined Based on Bahls (1992). Nonmetric Mann-Whitney U Test Was Used to Test Differences of Two Groups.

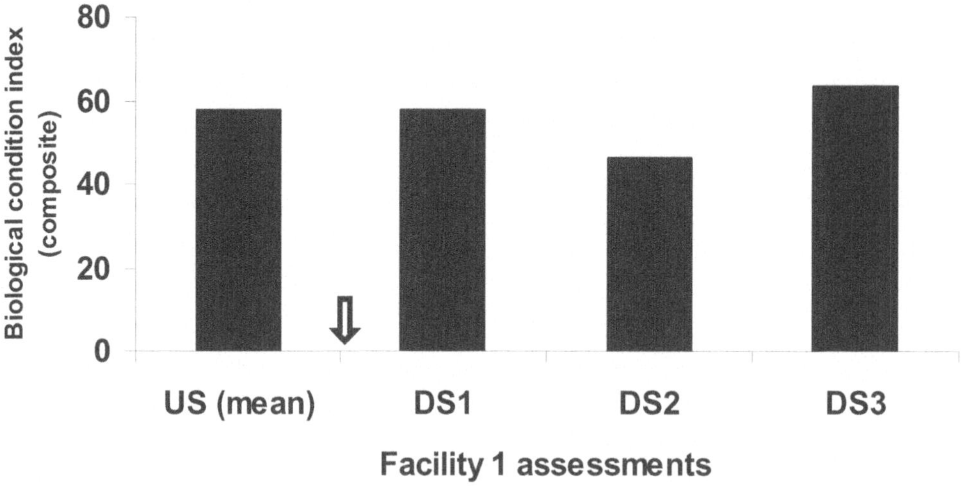

Figure 4-3. Biological Condition Composite Indexes for Facility 1 Based on Benthic Macroinvertebrate, D-frame Samples. Comparison Between Mean of Upstream (US) Samples (Mean 57.9, Distinguishable Difference at 90% Confidence (DD_{90}) =1.3 n=3) to Three Samples Progressively Further Downstream (DS) of the Effluent Discharge. Arrow Indicates Discharge Point.

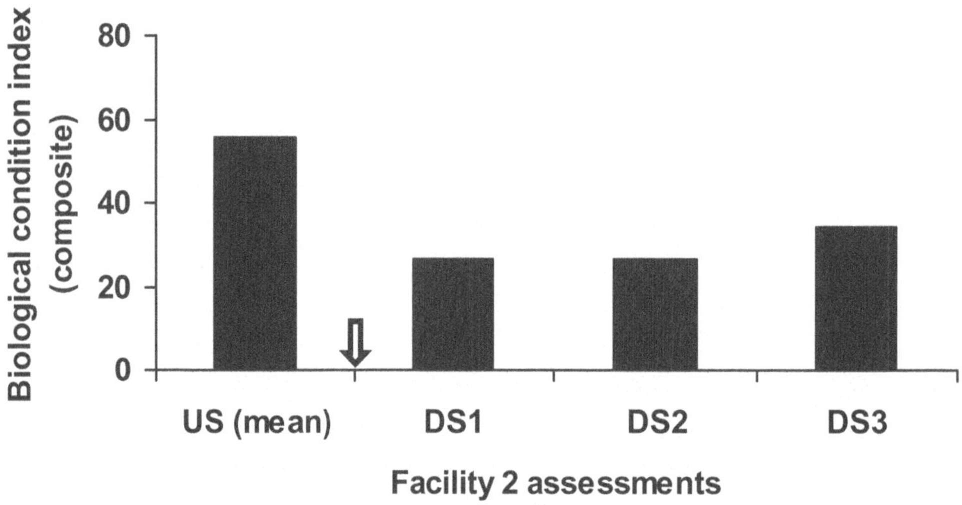

Figure 4-4. Biological Condition Composite Indexes for Facility 2 Based on Benthic Macroinvertebrate, D-frame Samples. Comparison Between Mean of Upstream (US) Samples (Mean 55.6, DD_{90}=1.0, n=3) to Three Samples Progressively Further Downstream (DS). Arrow Indicates Discharge Point.

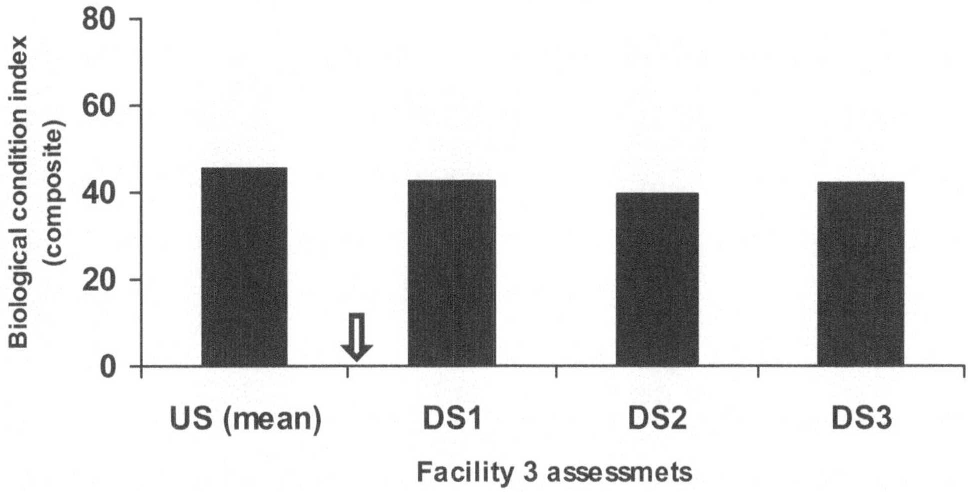

Figure 4-5. Biological Condition Composite Indexes for Facility 3 Based on Benthic Macroinvertebrate, D-frame Samples. Comparison Between Mean of Upstream (US) Samples (Mean 45.6, DD_{90}=1.3, n=3) to Three Samples Progressively Further Downstream (DS). Arrow Indicates Discharge Point.

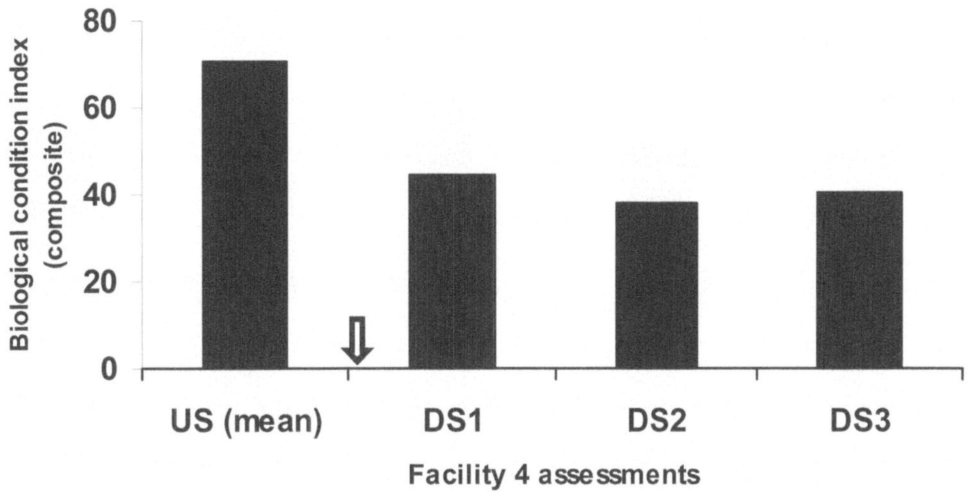

Figure 4-6. Biological Condition Composite Indexes for Facility 4 Based on Benthic Macroinvertebrate, D-frame Samples. Comparison Between Mean of Upstream (US) Samples (Mean 70.3, DD_{90}=0.6, n=3) to Three Samples Progressively Further Downstream (DS). Arrow Indicates Discharge Point.

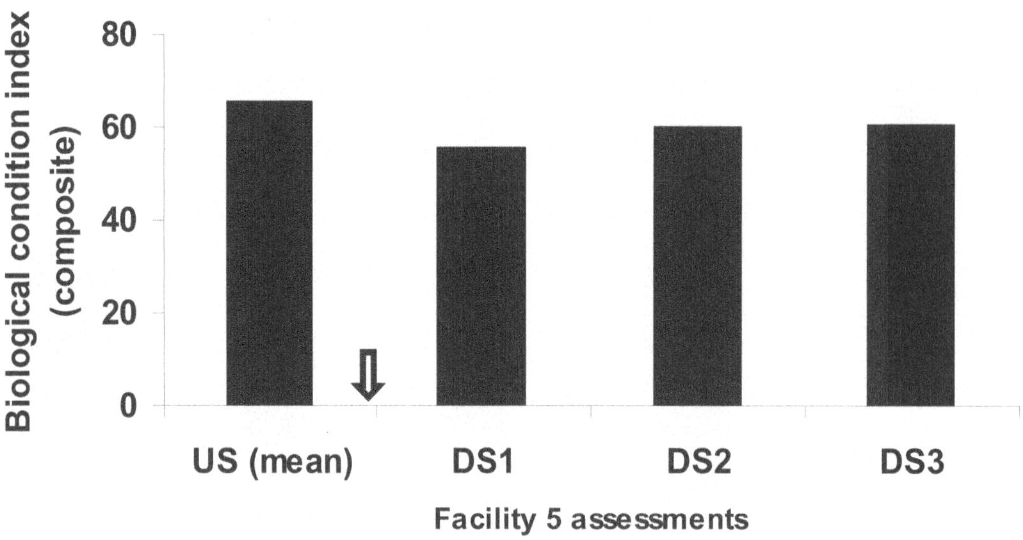

Figure 4-7. Biological Condition Composite Indexes for Facility 5 Based on Benthic Macroinvertebrate, D-frame Samples. Comparison Between Mean of Upstream (US) Samples (Mean 65.4, DD_{90}=1.4, n=3) to Three Samples Progressively Further Downstream (DS). Arrow Indicates Discharge Point.

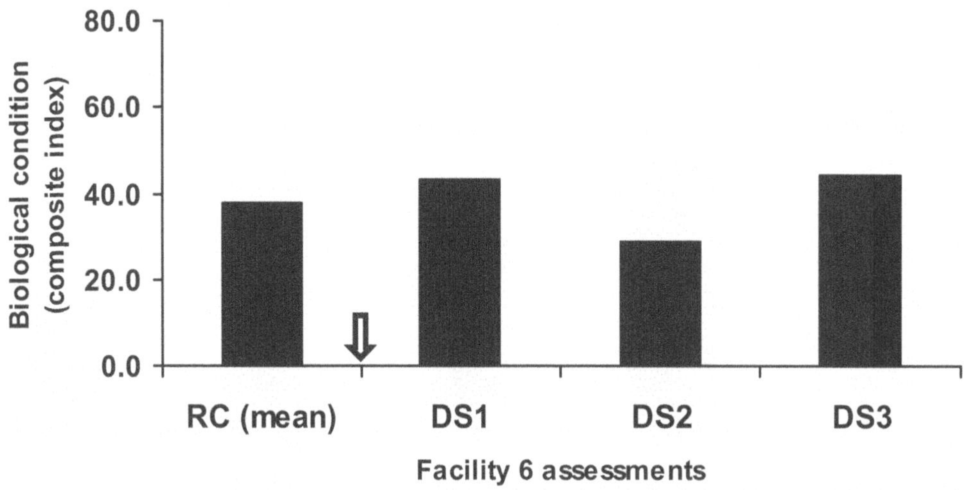

Figure 4-8. Biological Condition Scores for Facility 6 Based on Benthic Macroinvertebrate, D-frame Samples. Comparison Between Mean of Reference Condition (RC) Samples (Mean 37.8, DD90=4.5, n=2) to Three Samples Progressively Further Downstream (DS). Arrow Indicates Discharge Point.

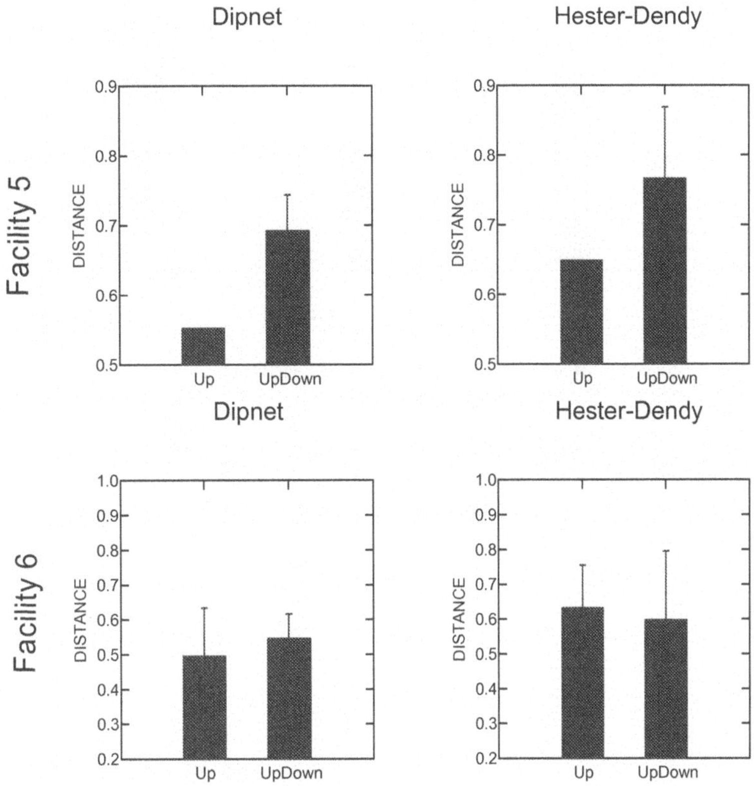

Figure 4-9. Comparison of Macroinvertebrate Assemblage Dissimilarity Between Hester-Dendy Artificial Substrate Samplers and the Multi-habitat Dip Net Sampling Method Used Up and Downstream of Facilities 5 and 6. Up = Similarity of Upstream Samples; upDown is Similarity Between up and Downstream Samples. Higher Distance Values Indicate Less Similarity in Assemblage Composition Between Samples.

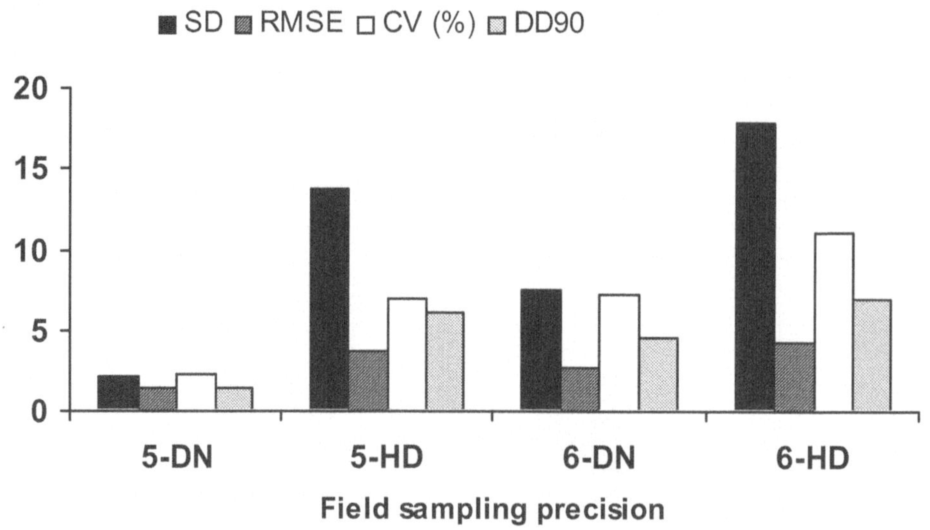

Figure 4-10. Comparison of Precision Estimates for Dipnet (DN) and Hester-Dendy (HD) Samples Taken at Facility 5 (n=3) and Facility 6 (n=2). Abbrev: SD (Standard Deviation), RMSE (Root Mean Squared Error), CV (Coefficient of Variation), and DD90 (Detectable Difference at 90% Confidence).

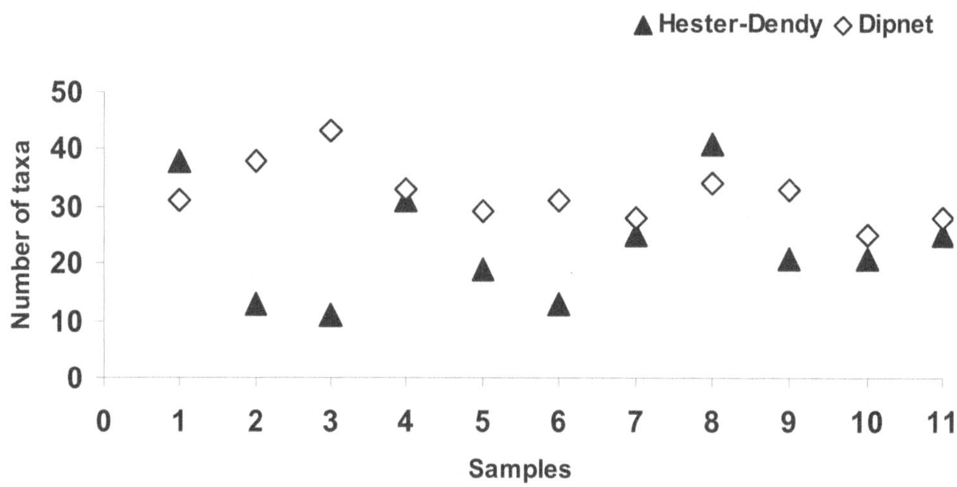

Figure 4-11. Sample by Sample Comparison of Metric "Number of Taxa" Between Two Sampling Methods.

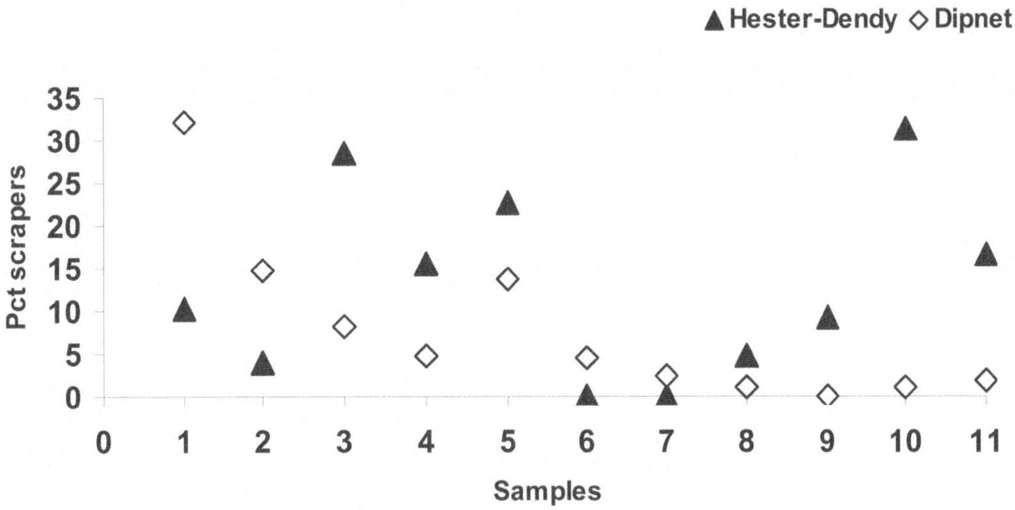

Figure 4-12. Sample by Sample Comparison of Metric "Percent Scrapers" Between Two Sampling Methods.

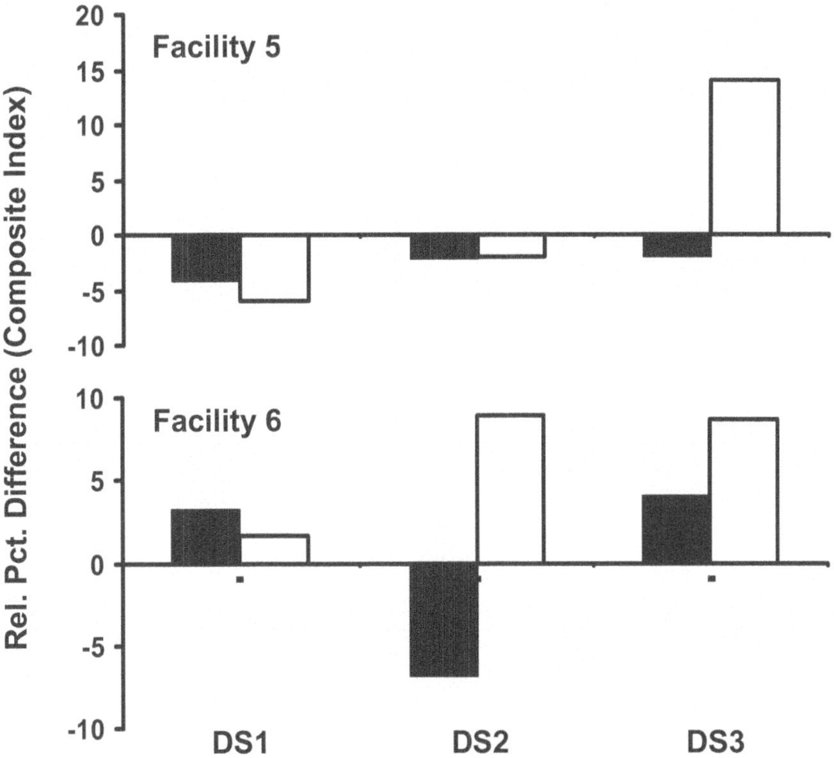

Figure 4-13. Relative Percent Difference (RPD) of Composite Biological Index Scores from Samples Taken Side by Side Using Field Gear. DS1-3 are Sites Sequentially Downstream from Discharge Point; Closed Bars are Dipnet Samples, and Open Bars are Hester-Dendy Samples.

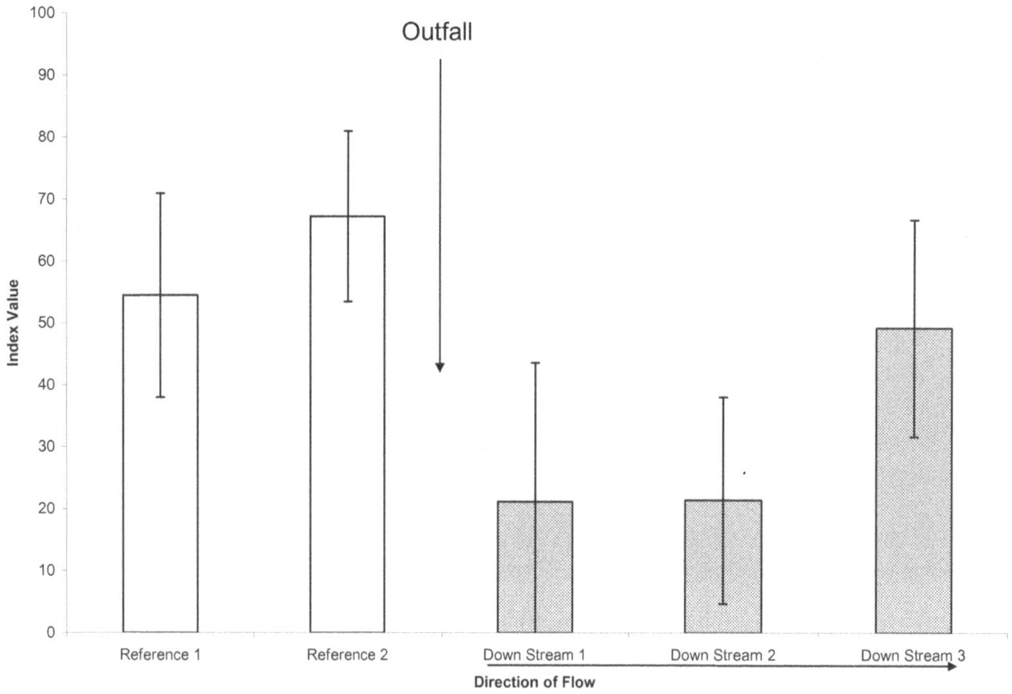

Figure 4-14. Fish index Score at Sites Up and Downstream of Facility 6. This Facility Did Not Have Viable Upstream Flow and Instead Relied on Two Reference Streams in the Region.

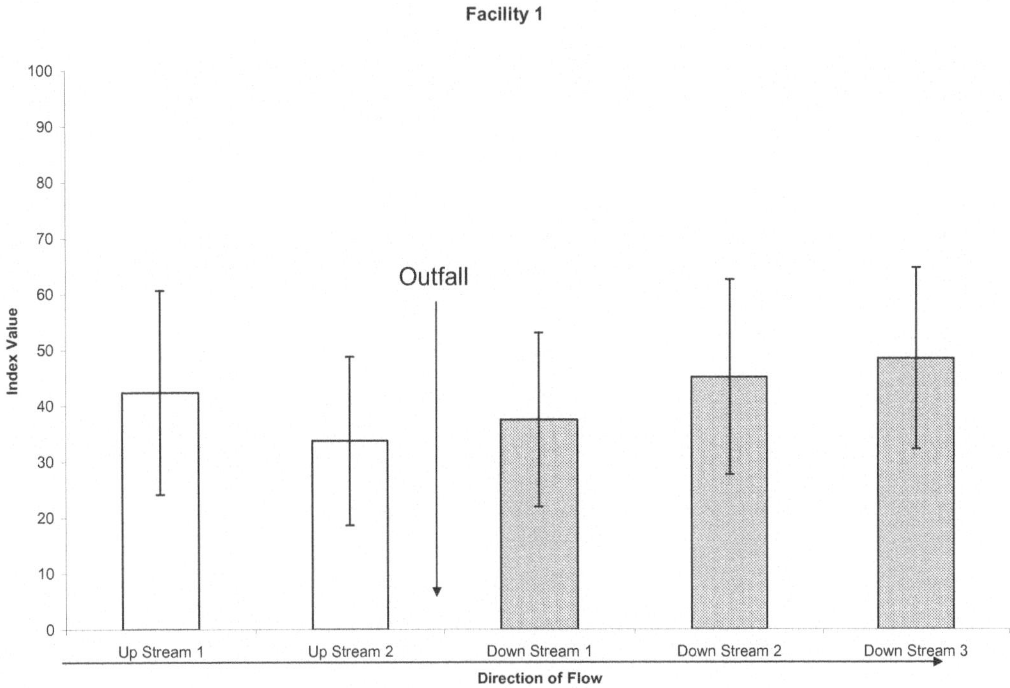

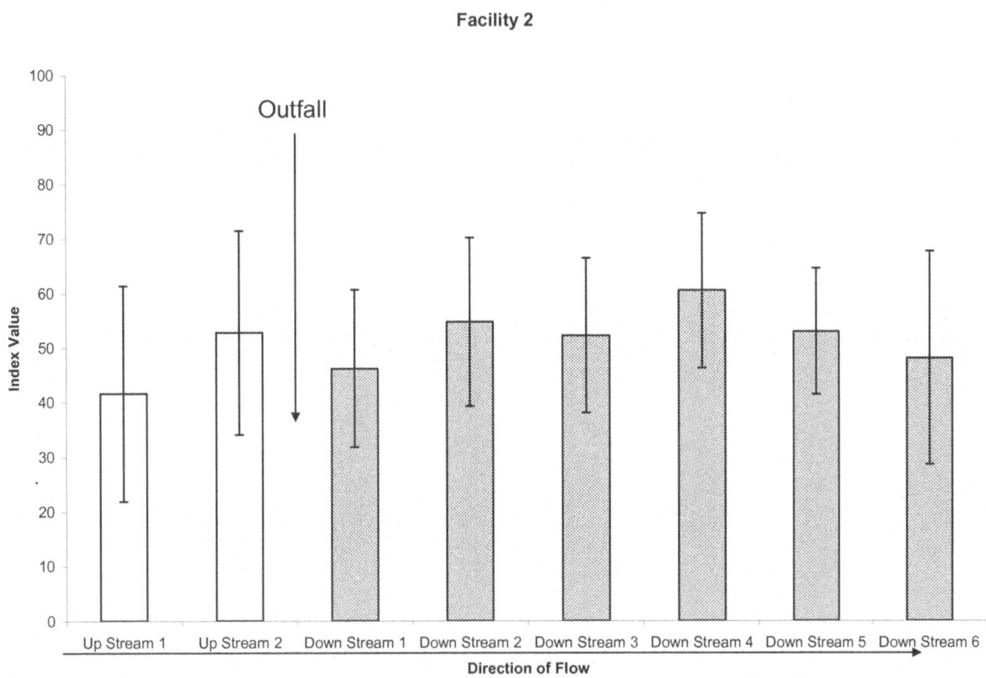

Figure 4-15. Fish Index Scores and 95% Confidence Intervals (Error Bars) Calculated for Upstream (Clear Bars) and Downstream (Shaded Bars) Locations Relative to Discharges from Facilities 1 and 2.

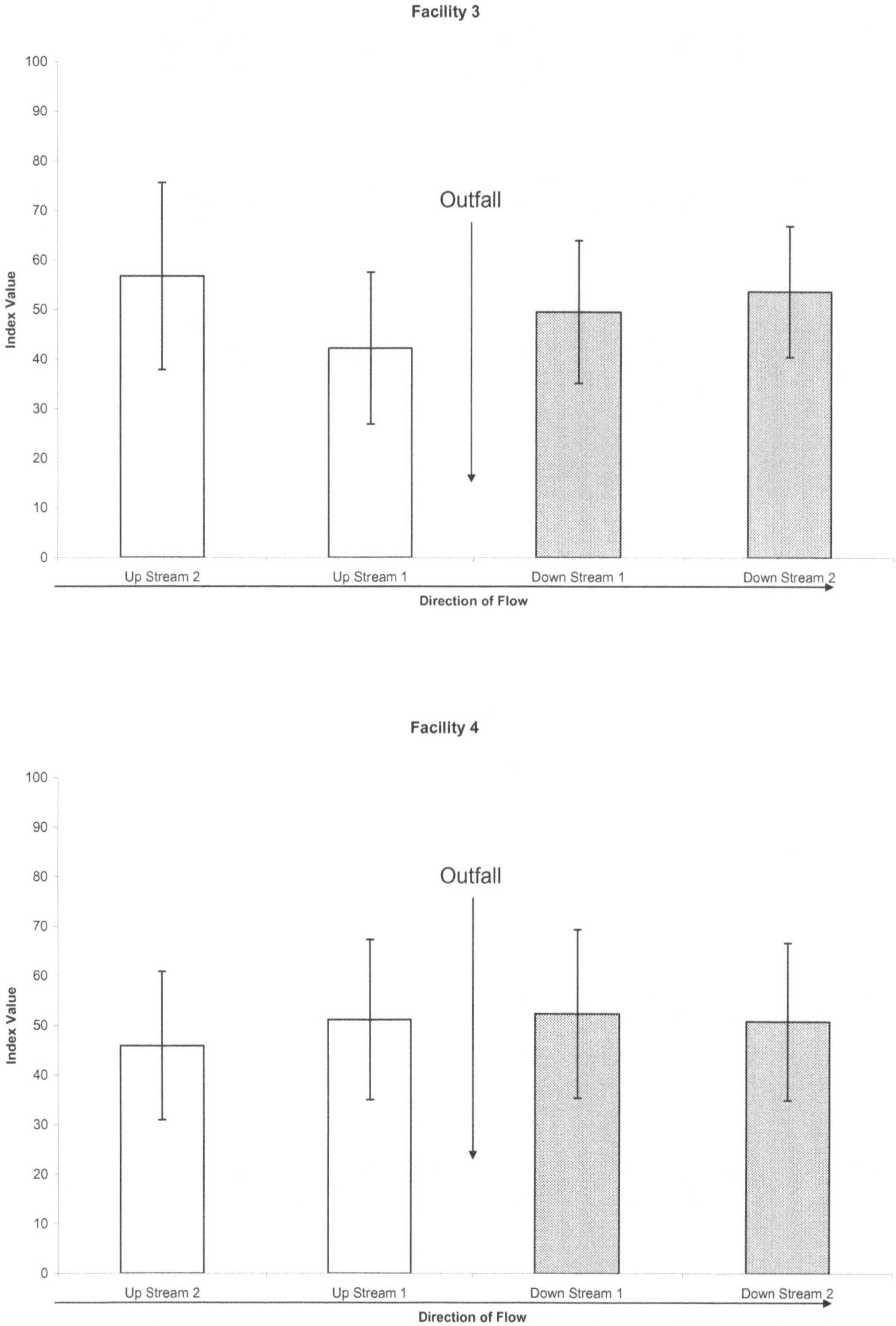

Figure 4-16. Fish Index Scores and 95% Confidence Intervals (Error Bars) Calculated for upstream (Clear Bars) and Downstream (Shaded Bars) Locations Relative to Discharges from Facilities 3 and 4.

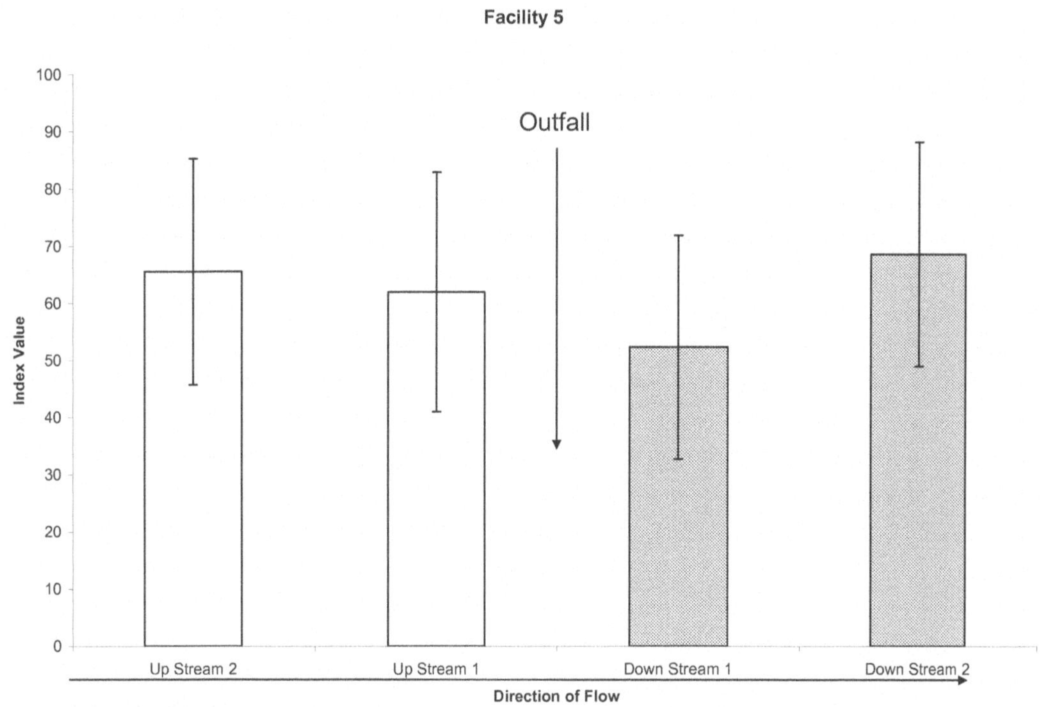

Figure 4-17. Fish Index Scores and 95% Confidence Intervals (Error Bars) Calculated for Upstream (Clear Bars) and Downstream (Shaded Bars) Locations Relative to Discharges from Facility 5.

CHAPTER 5.0

RELATIONSHIPS BETWEEN WET AND BIOASSESSMENT RESULTS

Table 5-1 summarizes the percent failed WET test results and corresponding bioassessment results by facility. As noted in Chapter 3.0, for most facilities, a higher percentage of tests failed the WET limit if the *S. capricornutum* results are included. Based on all WET tests completed at each facility, there appeared to be little or no relationship between the percentage of "failed" tests and bioassessment results overall (Table 5-2; $r = -0.27$, $p = 0.64$; Figure 5-1). Considering the periphyton bioassessment results, for example, four facilities having between 6.7% and 58.3% "failed" tests were all associated with some form of periphyton "impairment" downstream. Similarly, the facility with the lowest WET percent "failed" rate (6.7% for Facility 6) was the only one that exhibited significant changes in fish assemblage condition downstream.

There was little evidence of relationships considering specific WET test types and analogous bioassessment assemblages (Figure 5-2). *S. capricornutum* test results appeared to be unrelated to observed changes in periphyton assemblage downstream of facilities ($t = 0.67$, $p = 0.54$; Figure 5-2). Facility 6, which reported no algal toxicity in WET testing, exhibited a significant decrease in periphyton condition compared to reference sites while Facility 5, which reported high toxicity to *S. capricornutum*, exhibited no significant changes in periphyton assemblage condition.

Little evidence of relationships was also apparent based on either macroinvertebrates or fish (Figure 5-2). Facility 2, which had a "failed" test rate of 50% for the *P. promelas* test (based on NOECs) showed little change in the fish assemblage up and downstream while Facility 6, which failed one out of five fish tests, exhibited apparent impairment to the fish assemblage.

Magnitude of toxicity appeared to be somewhat correlated to bioassessment results. Using actual effluent dilution at the time of testing, Facility 6 exhibited the most toxicity (16 TU_c) in terms of either *C. dubia* or *P. promelas* tests while Facilities 1 and 5 had TU_cs all <1.0. The latter two facilities exhibited no observed detrimental biological changes downstream and Facility 6 exhibited significant changes in two of the three assemblages. Examining magnitude in terms of WET test type and the analogous biological assemblage yielded mixed results in terms of significant relationships. Facility 5 exhibited the highest toxicity to *S. capricornutum* (based on actual effluent dilution (14.4 TU_c) but detrimental changes were not observed downstream. Facility 1 exhibited fairly high toxicity to *C. dubia* and *P promelas* (2.5 and 2.4 TU_c, respectively) but did not have observed detrimental changes in either the macroinvertebrate or fish assemblages downstream.

Some relationships between WET and biological effects may have been observed at Facilities 2 and 6. Facility 2, which had the highest observed toxicity to *C. dubia* in this study (15 TU_c), also exhibited a significant decrease in the macroinvertebrate index downstream. Facility 6, which exhibited the highest toxicity to *P. promelas* in this study (15 TU_c based on NOEC), had a significant decrease in the fish assemblage index downstream. However, both of these facilities had perhaps the most striking differences between up and downstream habitats, with Facility 2 having a difference of about 50 MGD between up and downstream, and Facility 6 having reference sites and no upstream. Thus, a full study would need to have much more intensive sampling up and downstream and a more thorough understanding of reference condition for the downstream sites being examined.

It should be emphasized that the analyses presented above were based on changes in biological assemblages up and downstream of a given facility. To the extent that upstream sites are severely impaired due to other sources upstream, effects of an effluent on downstream sites would not be easily distinguishable using this analysis. As reference conditions were not sampled in this Pilot Study (except at Facility 6), upstream biological condition, in an absolute sense, could not be determined. This comparison might be useful in a full study to determine the degree of biological change that could be distinguished relative to upstream conditions. In this study, most facilities had relatively good quality upstream habitats and no known sources of toxicity within at least 3 km upstream of each facility. However, some fish metric results suggested only fair conditions upstream (e.g., Facility 2) based on typical reference conditions for the region, suggesting the need for more careful assessment approaches in a full study depending on the upstream conditions,.

Table 5-1. Summary of Whole Effluent Toxicity and Bioassessment Results from the Six Facilities. + - No Detrimental Change Downstream, – = Detrimental Change Observed Downstream. C.d. = *Ceriodaphnia dubia*; P.p. = *Pimephales promelas*; S.c. = *Selenastrum capricornutum*.

	Percent Tests "Failed"[1]					Bioassessment Results			
Facility	C.d.	P.p.	S.c.	Total	Total-Algae[2]	Periphyton	Macroinvertebrates	Fish	Total[3]
1	25	25	100	40	20	+	+	+	0
2	25	50	100	44	37.5	–	–	+	2
3	25	75	100	58.3	44.4	–	+	+	1
4	0	25	100	36.3	12.5	–	–	+	2
5	0	0	100	33.3	0	+	+	+	0
6	0	20	0	6.7	10.0	–	+	–	2

[1] "Failed" tests are those WET tests that reported endpoints not meeting the facility WET limit or trigger concentration under either design or actual flow conditions.

[2] Total percentage of only *C. dubia* and *P. promelas* tests that "failed".

[3] Total of communities with observed downstream detrimental change.

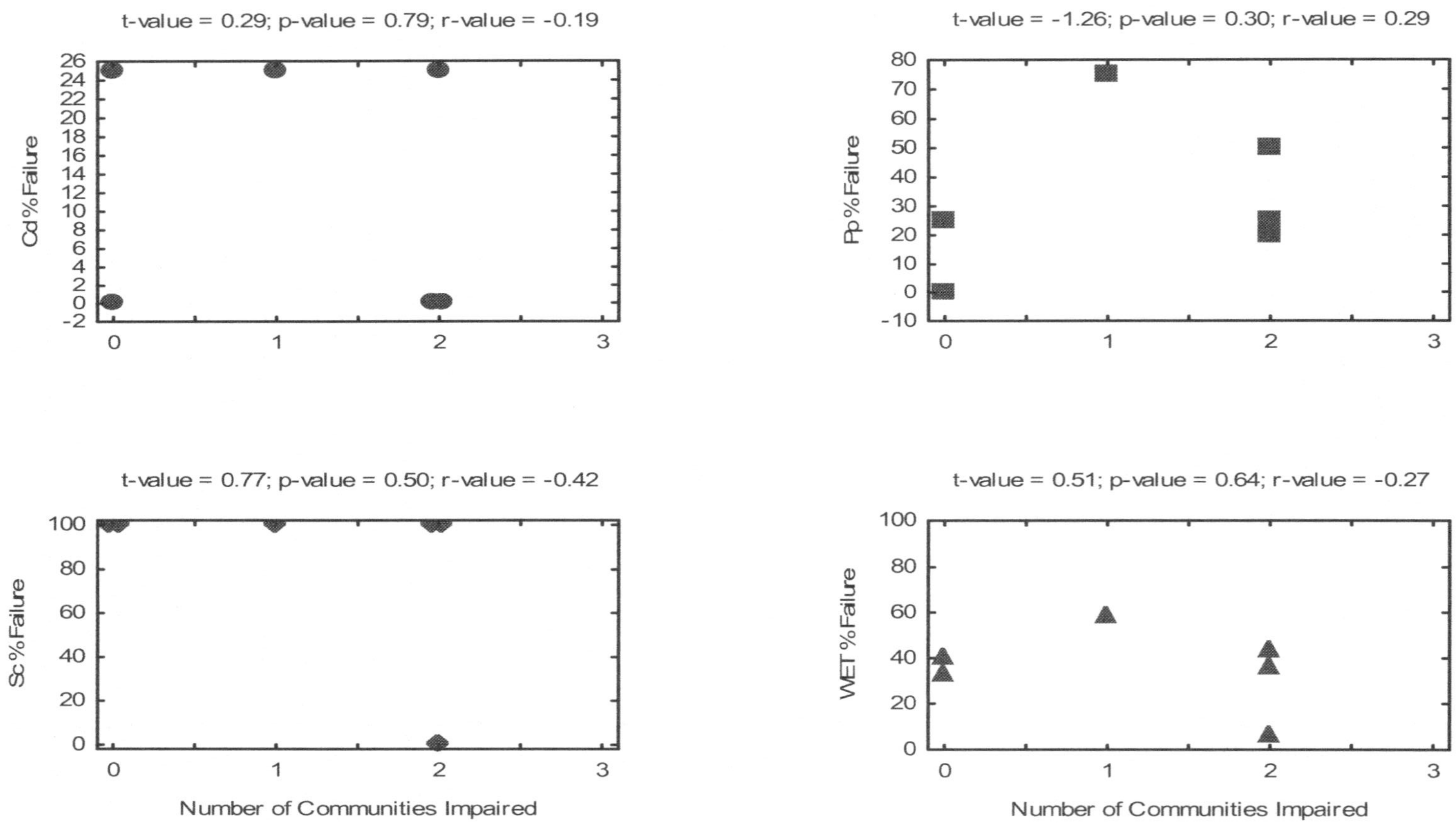

Figure 5-1. Correlations Between the Percentage of "Failed" WET Tests, by Test Type, and the Number of Assemblages for Which a Decrease in Biological Condition Downstream of a Facility Was Observed. Cd = *Ceriodaphnia dubia*, Pp = *Pimephales promelas*, and Sc = *Selenastrum capricornutum*. The Figure in the Lower Right Represents the Relationship Between the Total Percent "Failed" WET Tests and the Number of Assemblages for Which Effects Were Observed.

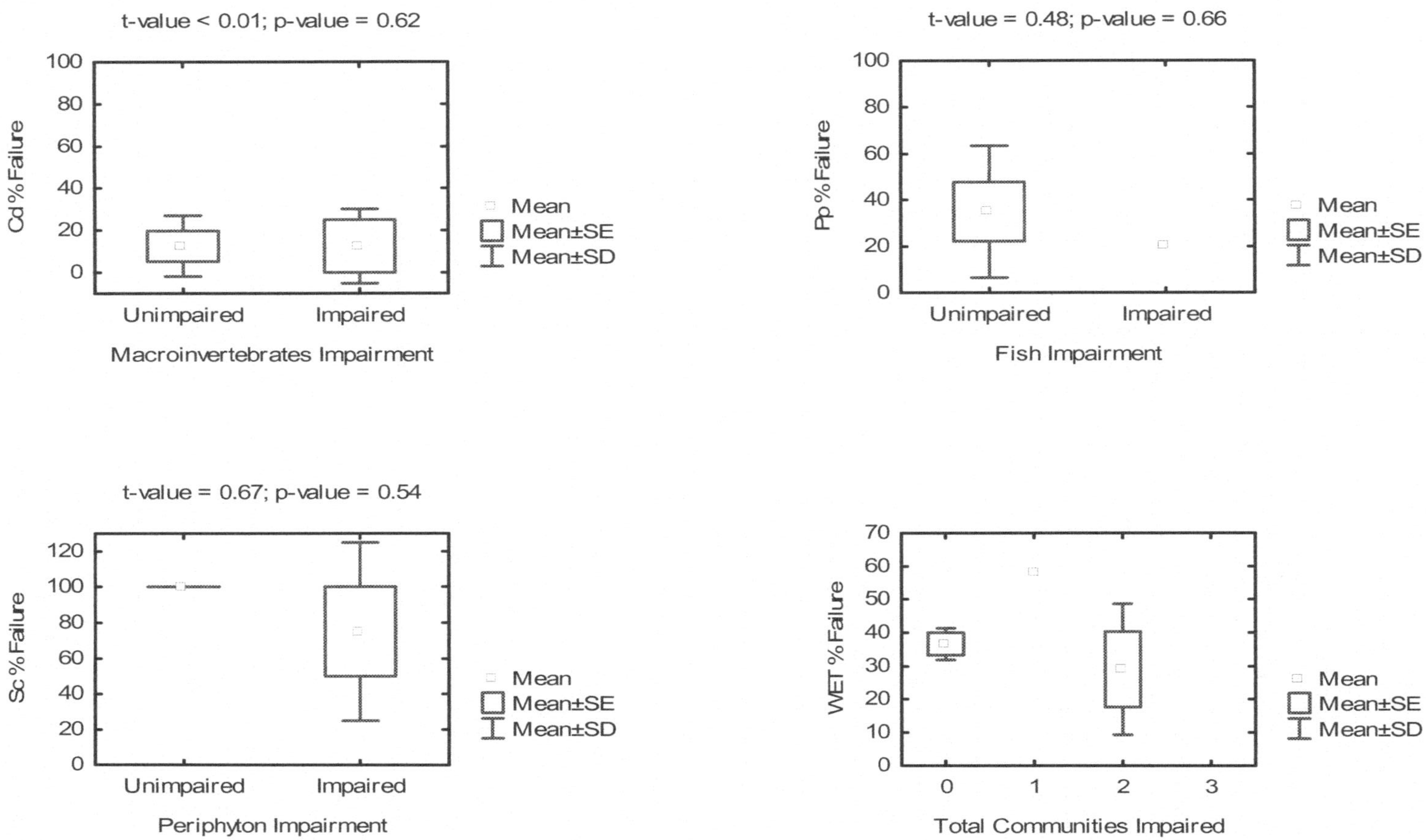

Figure 5-2. Box Plots Of The Percent "Failed" WET Tests, by Test Type, as a Function of the Number of Assemblage Types Showing Some Form of Decrease in Biological Condition Downstream of a Facility. See Legend for Figure 5-1 for Species Codes. The Figure in the Lower Right Represents the Relationship Between Total Percent WET Test "Fails" and the Number of Biological Assemblages Showing Detrimental Effects.

CHAPTER 6.0

CONCLUSIONS AND RECOMMENDATIONS

Results of this pilot study demonstrate the importance of obtaining high-quality WET and bioassessment data prior to making compliance determinations for a permitted wastewater effluent discharge. Although only six wastewater dischargers and their WET laboratories were involved in this study, it appears certain that compliance with WET test acceptance criteria alone will not guarantee high quality WET data. All of the QC testing in this Pilot suggests that compliance with test acceptance criteria, while necessary and important, is not sufficient for evaluating laboratory performance and data quality for freshwater chronic WET tests. Other quality control criteria (many of which are recommended by U.S. EPA as well as the Pellston Workshop on WET held in 1995 (Goethe et al., 1996)) are needed to ensure that correct test interpretations are made. These other criteria (or MQOs as termed in this study) include minimum and maximum percent minimum significant difference (PMSD), minimum accuracy thresholds as determined in routine testing of blind and positive control samples, and recognition of test-test endpoint variability and varying test organism sensitivity, as evidenced by reference toxicant test endpoint inter-test precision.

Many of the MQOs recommended in this study have been discussed in recent publications (Gulley et al., 2000; Shukla et al., 2000; U.S. EPA, 2002b). The suggested alternative WET analysis presented in Chapman et al. (1996), for example, (which requires meeting PMSD criteria prior to use and comparison with bioequivalence endpoints prior to interpretation; Shuckla et al., 2000) appears to be a useful start towards encouraging higher quality WET data. Using such a flow-chart, some of the disparity observed in this study in interpretations of WET results based on either NOECs or IC_{25}s would be resolved. Without such as analysis framework, this study suggests that interpretations of WET "pass" or "fail" are very dependent on which type of statistical endpoint is used. Results of this study demonstrate that both NOEC and IC_{25}s can yield uncertain assessments of WET for a given sample if acceptable test variability is not controlled via some form of defensible criteria. However, other data quality factors besides PMSDs are necessary to control false positive and negative rates of WET tests. The green algae WET test (*Selenastrum capricornutum*) results appeared suspect in this study due to a high false positive rate and poor agreement in split testing, yet these tests consistently met all acceptance criteria. While these results might be a function of the laboratories involved in this study, further evaluation of the four-day freshwater algal WET test, or the implementation of the test, is warranted.

This study also showed that bioassessment data needs to be of high quality to ensure accurate decisions regarding effects of an effluent discharge. While lab processing and field replication generally met MQOs for periphyton and macroinvertebrate data, other MQOs, such as calibrated indicators for the ecoregion of interest (Barbour et al., 1999) and adequate bioassessment sensitivity need further attention (Diamond et al., 1996). Indeed, many state programs have begun to characterize and refine their bioassessment programs, which promises to improve the rigor and sensitivity of these methods. Fish bioassessments were less precise in general than either periphyton or macroinvertebrate assessments and they require more field effort. However, fish assessments in general have routinely proved to be extremely informative (Yoder and DeShon, 2003; Emery and Thomas, 2003) and form the backbone of many monitoring programs. For the purposes of relating WET to instream biological condition, though, it appears that both macroinvertebrate and periphyton assessments at a site will address impacts that might occur to fish as well. Notwithstanding the preliminary nature of these pilot results, periphyton and macroinvertebrate assessments were more effective then fish in indicating an impact.

Comparison of artificial substrate and multi-habitat dipnet macroinvertebrate sampling suggested similar results at the two site examined. However, the dipnet method yielded slightly better performance based on the samples analyzed and may be a better choice in many cases due to the extra field time and expense needed for artificial substrate sampling. In some sites, though, dipnet sampling may be impractical, or habitat may be of too poor quality, making artificial substrate samples more necessary.

This study demonstrated ways in which WET results could be compared with biological condition but the results observed here should not be considered to be a definitive assessment of relationships between the two types of monitoring. This was a pilot study and so was not designed to determine definitive relationships between WET and biological condition. Certain factors examined in this study, such as type of statistical endpoint (NOEC vs IC_{25}), magnitude of toxicity versus frequency of WET limit exceedence, and use of design low flow versus actual effluent dilution in interpreting WET results, all appear to have some influence on relationships observed (or not) with biological conditions in this study. These factors should be examined in future studies to identify ways to more accurately interpret WET test results and to improve WET test implementation in NPDES permits.

Recommendations for Comprehensive Study

The overall goal of this pilot study was to determine how a comprehensive study should be designed to adequately address relationships between WET and instream biological condition. The pilot study was able to further define the methods measurement quality objectives (MQOs) to be used in a more definitive study as summarized below and in Table 6-1.

All proposed MQOs for WET and bioassessments, tested in this pilot study, were shown to be achieveable and are necessary to ensure high quality data. This project indicated that WET testing acceptability criteria are not enough to provide the high quality data that is required to relate WET to instream biological changes. Additional WET criteria, including minimum and maximum percent significant differences, routine blind positive and negative control samples, and some split testing are required to ensure high quality WET data. These additional QC WET tests need to be conducted at least once in a comprehensive study. Results of the pilot study also indicated that only the 7-d *Ceriodaphnia* and *Pimephales* short-term chronic tests should be conducted. The *Selenastrum* 4-d test was associated with too much uncertainty in this study and it is doubtful that a sufficient number of laboratories are available to routinely conduct this test on a more widespread basis. Both NOECs and IC_{25} should be calculated for each WET test, in addition to using a WET percent effect statistical approach (e.g., Shukla et al., 2000; Chapman et al., 1996). Also, any laboratory involved in WET testing in the definitive study must conduct monthly chronic reference toxicant tests using either sodium chloride or potassium chloride. Finally, quarterly WET testing should be considered a minimum for a comprehensive study. Monthly testing would provide more precision in the data interpretations.

Based on the results of the biological assessments conducted in the pilot, fish bioassessments were less precise in general than either periphyton or macroinvertebrate assessments in indicating a change downstream of a discharge and is therefore not necessary. Only the macroinvertebrate and periphyton bioassessments are necessary. The dipnet method yielded better performance in comparison to the Hester-Dendy samplers for the assessment of macroinvertebrate community based on the analysis of samples. This protocol is also easier to conduct and less susceptible to data loss. Therefore, macroinvertebrates should be sampled via dipnet using a multi-habitat approach as done in this pilot, in the definitive study.

A definitive study should encompass at least two field bioassessments, spanning a total of two years of data collection (e.g., two summer-fall bioassessments at each site). This would entail two years of WET test data as well for each site, which would improve the ability to detect relationships between WET and bioassessments, if they occur.

The pilot study involved only six facilities, which was known to be a small sample size. Based on the MQOs identified and the various site-specific issues that were observed in the pilot, it would be desirable to have at least 50 sites, representing different regions of the U.S. Sites should also represent a range of design effluent dilutions, however, all sites would need to be wadeable to ensure adequate performance of the bioassessment protocols. In addition, it is critical that several sites be located in areas where there is no viable upstream and appropriate reference sites are not available. This is a common issue in western U.S. but was not an issue that needed to be evaluated in the pilot study.

It should be noted that the logistics of a more definitive study would be exponentially greater than those of the pilot study. For the most part the six facilities that were involved in the pilot study were very cooperative, but increasing the number of facilities an order of magnitude or greater for a more definitive study would also increase the issues surrounding scheduling, quality assurance, and financial obligations.

Table 6-1. Summary of Measurement Quality Objectives to be Used in a More Definitive Study.

Measurement Parameter	Measurement Quality Objectives
WET	
[1]Control survival (*C. dubia* and *P. promelas*)	≥ 80%
[1]Mean individual fish weight after a 7-day period	≥ 0.25 mg dry weight
[1]*C. dubia* offspring per adult female	≥ 15 within 7 days; ≥ 60 % controls have ≥ 3 broods within 7 days
[2]Percent Minimum significant difference (PMSD) for *P. promelas* control growth (intra-laboratory)	≤ 35%
[2]PMSD for *C. dubia* chronic control reproduction	≤ 37%
[3]CV for monthly reference toxicant test endpoints	≤ 35%
[3]Split testing of effluent sample; CV for endpoints	≤ 35%
[4]Split testing of method blank or positive control between labs: CV of endpoints	≤ 35%
[4]Associated reference toxicant test yields endpoint of the long-term average	± 35%
Bioassessment	
Field Sampling	
Precision of replicate samples	RPD ≤ 15%
Representativeness	Multi-habitat sampling
Macroinvertebrate Sorting/Subsampling	
Precision of composition metrics	RPD ≤ 15%
Accuracy of sorting	SE ≥ 90%
Algae Subsampling	
Accuracy of aliquot (for composition metrics between aliquot and total sample)	RPD ≤ 5%
Taxonomy	
Precision of identified individuals	PTD ≤ 12%

[1] Groethe, et al., 1996; U.S. EPA, 2000; U.S. EPA, 2002a.
[2] U.S. EPA, 2002b.
[3] Groethe, et al., 1996; U.S. EPA, 2000; Warren-Hicks, et al., 2000; Moore, et al., 2000.
[4] U.S. EPA, 2000b; Moore, et al., 2000.

REFERENCES

Ankley, G. 1997. Laboratory vs field measurement endpoints: a contaminated sediment perspective. In: C. Ingersoll et al. (eds) Ecological Risk Assessment of Contaminated Sediments. Society of Environmental Toxicology and Chemistry (SETAC) Press, Pensacola, FL., pp.115-122.

ASTM. 2000. E729-96 Standard guide for conducting acute toxicity tests with fishes, macroinvertebrates, and amphipods. pp. 218-238. *In*: 1998 Annual Book of ASTM Standards, Vol. 11.05, Philadelphia, PA.

Baird, R., R. Berger, and J. Gully, 1996. Improvements in point estimation methods and application to controlling aquatic toxicity test reliability. In D. Groethe, K. Dickson, and D. Reed-Judkins. *Whole effluent toxicity testing: An evaluation of methods and prediction of receiving system impacts*, Society of Environmental Toxicology and Chemistry, SETAC Press, Pensacola, FL. pp. 103-130.

Barbour, M.T., J.B. Stribling, and P.F.M. Verdonschot. 2006. The multihabitat approach of U.S. EPA's rapid bioassessment protocols: Benthic macroinvertebrates. Limnetica (in press).

Barbour, M.T., J. Gerritsen, B.D. Snyder, and J.B. Stribling. 1999. Rapid Bioassessment Protocols for Use in Streams and Wadeable Rivers: Periphyton, Benthic Macroinvertebrates and Fish, Second Edition. EPA 841-B-99-002. U.S. Environmental Protection Agency; Office of Water; Washington, D.C.

Barbour, M.T., J.M. Diamond, C.O. Yoder. 1996. Biological assessment strategies: Applications and Limitations. Pages 245-270 *in* D.R. Grothe, K.L. Dickson, and D.K. Reed-Judkins (editors). *Whole effluent toxicity testing: An evaluation of methods and prediction of receiving system impacts*, Society of Environmental Toxicology and Chemistry, SETAC Press, Pensacola, FL. pp. 245-270.

Chapman, G., et al. 1996. Methods and Appropriate Endpoints. *in*: D.R. Grothe, K.L. Dickson, and D.K. Reed-Judkins (editors). *Whole effluent toxicity testing: An evaluation of methods and prediction of receiving system impacts*, SETAC Press, Pensacola, FL. pp. 51-82.

Diamond, J. and C. Daley. 2000. What is the relationship between whole effluent toxicity result and biological condition? *Environ. Toxicol. Chem.* 19:158-168.

Diamond, J., C. Daley, T. Moore. 1999. Evaluating Whole Effluent Toxicity Testing as an Indicator of Instream Biological Conditions. Water Environment Research Foundation Report, 95-HHE-1, Alexandria, VA.

Diamond, J., J. Stribling, and C. Marino. 2002. Draft Study Design: Comparing WET to Biological Assessments. Project 00-ECO-5. Water Environment Research Foundation, Alexandria, VA.

Diamond, J., J.B. Stribling, and M.T. Barbour. 1996. Characterizing and comparing bioassessment approaches and their results: A perspective. *J. North American Benthological Society* 15:713-727.

Dickson, K.L., W.T. Waller, J.H. Kennedy, and L.P. Ammann. (1992). Assessing the relationship between ambient toxicity and instream biological response. *Environ. Toxicol. Chem.* 11:1307-1322.

Dickson, K.L., W.T. Waller, J.H. Kennedy, L.P. Ammann, R. Guinn, and T.J. Norberg-King. (1996). Relationships between effluent toxicity, ambient toxicity, and receiving system impacts: Trinity River dechlorination case study. *In* Grothe, D.R., K.L. Dickson, and D.K. Reed-Judkins (eds.). *Whole Effluent Toxicity Testing: An Evaluation of Methods and Prediction of Receiving Stream Impacts*. Society of Environmental Toxicology and Chemistry, SETAC Press, Pensacola, FL. pp. 287-308.

Eagleson, K., D. Lenat, L. Ausley, and F. Winborne. (1990). Comparison of measured instream biological responses with responses predicted by *Ceriodaphnia* chronic toxicity tests. *Environ. Toxicol. Chem.* 9:1019-1028.

Emery, E. and J. Thomas. 2003. A method for assessing outfall effects on great river fish populations: the traveling zone approach. In: T. Simon (ed) Biological Response Signatures: Indicator patterns using aquatic communities. CRC Press, Boca Raton, FL, pp. 157-164.

Gallardo, A.C., J.E. Garrish, C.R. Hill, and J.B. Stribling 2006. Biological Assessment of the Cattail Creek and Brighton Dam Watershed, Howard County, Maryland. Prepared by Tetra Tech, Inc., Owings Mills, MD for Howard County, Department of Public Works. Stormwater Management Division. Columbia, MD. January 2006. (for further information, contact Ms. Angela Morales at 410-313-6586, or by email amorales@co.ho.md.us).

Grothe, D.R., K.L. Dickson, and D.K. Reed-Judkins (editors) 1996. *Whole effluent toxicity testing: An evaluation of methods and prediction of receiving system impacts*, SETAC Press, Pensacola, FL.

Gulley, J., R. Baird, P. Markle, and J. Bottomley, 2000. Effect-based interpretation of toxicity test data using probability and comparison with alternative methods of analysis. Environ. Tox. Chem 19: 133-140.

Hill, C.R., J.B. Stribling, and A.C. Gallardo. 2005. *Documentation of Method Performance Characteristics for the Anne Arundel County Biological Monitoring Program*. Prepared by Tetra Tech, Inc., Owings Mills, MD for Anne Arundel County Office of Environmental & Cultural Resources. Annapolis, MD.

Karr, J. and E. Chu. 1999. Restoring life in running waters. Island Press, Washington, D.C.

La Point, T.W. and W.T. Waller. 2000. Field assessments in conjunction with whole effluent toxicity testing. *Environ Toxicol Chem* 19:14-24.

Marcus, M., L. McDonald. 1992. Evaluating the statistical bases for relating receiving water impacts to effluent and ambient toxicities. *Environ. Toxicol. Chem.* 11:1389-1402.

McCune B.M. and J.B. Grace. 2002. Analysis of Ecological Communities. MjM Software Design, Gleneden Beach, OR.

Moore, T., S. Canton, and M Grimes. 2000. Investigating the incidence of Type 1 errors for chronic whole effluent toxicity testing using *Ceriodaphnia dubia. Environ. Toxicol. Chem.* 19: 118-122.

Rosenberg, D.M., and V.H. Resh. 1982. The Use of Artificial Substrates in the Study of

Freshwater Benthic Macroinvertebrates. Chapter 6, *in* John Cairns, Jr. (editor), Artificial Substrates. Ann Arbor Science Publisher, Inc. Ann Arbor, MI. ISBN 0-250-40404-4.

Shukla, R., Q. Wang, F. Fulk, C. Deng, and D. Denton, 2000. Bioequivalence approach for whole effluent toxicity testing. Environ. Tox. Chem. 19: 169-174.

Stribling, J.B. , SR. Moulton, II, and G.L. Lester. 2003. Determining the quality of taxonomic data. Journal of the North American Benthological Society 22(4): 621-631.

Taylor, J.K. 1988. Defining the Accuracy, Precision, and Confidence Limits of Sample Data. Chapter 6, pages 101-107, *IN*, L. H. Keith (editor), *Principles of Environmental Sampling*. ACS Professional Reference Book. American Chemical Society. Washington, DC.

U.S. EPA. 2002a. Short-term methods for estimating the chronic toxicity of effluents and receiving waters to freshwater organisms, fourth edition, October 2002. EPA-821-R-02-013. Office of Water, Washington, D.C.

U.S. EPA. 2002b Understanding and Accounting for Method Variability in Whole Effluent Toxicity Applications Under the NPDES Program. EPA 833-R-00-003. Office of Water, Washington, D.C.

U.S. EPA. 2000. Preliminary report: Interlaboratory variability study of EPA short-term chronic and acute whole effluent toxicity test methods. EPA-821-R-00-028A, Office of Water, Washington, D.C.

U.S. EPA.1991. Technical support document for water quality-based toxics control. EPA-505/2-90-001, Washington, D.C.

Waller, W.T. et al. 1996. Predicting instream effects from WET tests. In: Grothe, D.R., Dickson K.L., Reed-Judkins, D.K., eds, *Whole Effluent Toxicity Testing: An Evaluation of Methods and Prediction of Receiving System Impacts.* Society of Environmental Toxicology and Chemistry, SETAC Press, Pensacola, FL. pp 271-286.

Warren-Hicks, W.J., and B.R. Parkhurst. 1996. Issues in whole effluent toxicity test uncertainty analyses. Discussion-Initiation Paper 6.3 (pages 180-189) In: Grothe, D.R., Dickson, K.L., Reed-Judkins, D.K.. *Whole effluent toxicity testing: an evaluation of methods and prediction of receiving system impacts.* Society of Environmental Toxicology and Chemistry, SETAC Press, Pensacola, FL. pp. 180-190.

Warren-Hicks, W., B. Parkhurst, D. Moore, R. Teed, R. Baird, R. Berger, D. Denton, J. Pletl, 2000. Assessment of whole effluent toxicity test variability: partitioning sources of variability. Environ. Tox. Chem 19: pp. 94-104.

Yoder, C. and J. DeShon. 2003. Using biological response signatures within a framework of multiple indicators to assess and diagnose causes and sources of impairments to aquatic assemblages in selected Ohio rivers and streams. In: T. Simon (ed) Biological Response Signatures: Indicator patterns using aquatic communities. CRC Press, Boca Raton, FL, pp. 83-96.

WERF SUBSCRIBERS

WASTEWATER UTILITY

Alabama
Montgomery Water Works & Sanitary Sewer Board

Alaska
Anchorage Water & Wastewater Utility

Arizona
Glendale, City of, Utilities Department
Mesa, City of
Peoria, City of
Phoenix Water Services Department
Pima County Wastewater Management
Safford, City of

Arkansas
Little Rock Wastewater Utility

California
Calaveras County Water District
Central Contra Costa Sanitary District
Corona, City of
Crestline Sanitation District
Delta Diablo Sanitation District
Dublin San Ramon Services District
East Bay Dischargers Authority
East Bay Municipal Utility District
Eastern Municipal Water District
El Dorado Irrigation District
Fairfield-Suisun Sewer District
Fresno Department of Public Utilities
Inland Empire Utilities Agency
Irvine Ranch Water District
Las Virgenes Municipal Water District
Livermore, City of
Los Angeles, City of
Los Angeles County, Sanitation Districts of
Napa Sanitation District
Orange County Sanitation District
Palo Alto, City of
Riverside, City of
Sacramento Regional County Sanitation District
San Diego Metropolitan Wastewater Department, City of
San Francisco, City & County of
San Jose, City of
Santa Barbara, City of
Santa Cruz, City of
Santa Rosa, City of
South Bayside System Authority
South Coast Water District
South Orange County Wastewater Authority
Stege Sanitary District
Sunnyvale, City of
Union Sanitary District
West Valley Sanitation District

Colorado
Aurora, City of
Boulder, City of
Greeley, City of
Littleton/Englewood Water Pollution Control Plant
Metro Wastewater Reclamation District, Denver

Connecticut
Greater New Haven WPCA

District of Columbia
District of Columbia Water & Sewer Authority

Florida
Broward, County of
Fort Lauderdale, City of
Miami-Dade Water & Sewer Authority
Orange County Utilities Department
Reedy Creek Improvement District
Seminole County Environmental Services
St. Petersburg, City of
Tallahassee, City of
Tampa, City of
Toho Water Authority
West Palm Beach, City of

Georgia
Atlanta Department of Watershed Management
Augusta, City of
Clayton County Water Authority
Cobb County Water System
Columbus Water Works
Fulton County
Gwinnett County Department of Public Utilities
Savannah, City of

Hawaii
Honolulu, City & County of

Idaho
Boise, City of

Illinois
American Bottoms Wastewater Treatment Plant
Greater Peoria Sanitary District
Kankakee River Metropolitan Agency
Metropolitan Water Reclamation District of Greater Chicago
Wheaton Sanitary District

Iowa
Ames, City of
Cedar Rapids Wastewater Facility
Des Moines, City of
Iowa City

Kansas
Johnson County Unified Wastewater Districts
Unified Government of Wyandotte County/ Kansas City, City of

Kentucky
Louisville & Jefferson County Metropolitan Sewer District
Sanitation District No. 1

Louisiana
Sewerage & Water Board of New Orleans

Maine
Bangor, City of
Portland Water District

Maryland
Anne Arundel County Bureau of Utility Operations
Howard County Department of Public Works
Washington Suburban Sanitary Commission

Massachusetts
Boston Water & Sewer Commission
Upper Blackstone Water Pollution Abatement District

Michigan
Ann Arbor, City of
Detroit, City of
Holland Board of Public Works
Saginaw, City of
Wayne County Department of Environment
Wyoming, City of

Minnesota
Rochester, City of
Western Lake Superior Sanitary District

Missouri
Independence, City of
Kansas City Missouri Water Services Department
Little Blue Valley Sewer District
Metropolitan St. Louis Sewer District

Nebraska
Lincoln Wastewater System

Nevada
Henderson, City of
Reno, City of

New Jersey
Bergen County Utilities Authority
Ocean, County of
Passaic Valley Sewerage Commissioners

New York
New York City Department of Environmental Protection

North Carolina
Charlotte/Mecklenburg Utilities
Durham, City of
Metropolitan Sewerage District of Buncombe County
Orange Water & Sewer Authority

Ohio
Akron, City of
Butler County Department of Environmental Services
Columbus, City of
Metropolitan Sewer District of Greater Cincinnati
Northeast Ohio Regional Sewer District
Summit, County of

Oklahoma
Oklahoma City Water & Wastewater Utility Department
Tulsa, City of

Oregon
Clean Water Services
Eugene, City of
Gresham, City of
Portland, City of
Water Environment Services

Pennsylvania
Philadelphia, City of
University Area Joint Authority

South Carolina
Charleston Water System
Mount Pleasant Waterworks & Sewer Commission
Spartanburg Water

Tennessee
Cleveland, City of
Knoxville Utilities Board
Murfreesboro Water & Sewer Department
Nashville Metro Water Services

Texas
Austin, City of
Dallas Water Utilities
Denton, City of
El Paso Water Utilities
Fort Worth, City of
Houston, City of
San Antonio Water System
Trinity River Authority

Utah
Salt Lake City Corporation

Virginia
Alexandria Sanitation Authority
Arlington, County of

Fairfax County
Hampton Roads Sanitation District
Henrico, County of
Hopewell Regional Wastewater Treatment Facility
Loudoun County Sanitation Authority
Lynchburg Regional WWTP
Prince William County Service Authority
Richmond, City of
Rivanna Water & Sewer Authority

Washington
Everett, City of
King County Department of Natural Resources
Seattle Public Utilities
Sunnyside, Port of
Yakima, City of

Wisconsin
Green Bay Metro Sewerage District
Kenosha Water Utility
Madison Metropolitan Sewerage District
Milwaukee Metropolitan Sewerage District
Racine, City of
Sheboygan Regional Wastewater Treatment
Wausau Water Works

Australia
South Australian Water Corporation
Sydney Water Corporation
Water Corporation of Western Australia

Canada
Regina, City of, Saskatchewan
Toronto, City of, Ontario
Winnipeg, City of, Manitoba

New Zealand
Watercare Services Limited

United Kingdom
Yorkshire Water Services Limited

STORMWATER UTILITY

California
Fresno Metropolitan Flood Control District
Los Angeles, City of, Department of Public Works
Monterey, City of
Sacramento, County of
San Francisco, City & County of
Santa Rosa, City of
Sunnyvale, City of

Colorado
Aurora, City of
Boulder, City of

Georgia
Griffin, City of

Iowa
Cedar Rapids Wastewater Facility
Des Moines, City of

Kansas
Overland Park, City of

Kentucky
Louisville & Jefferson County Metropolitan Sewer District

Maine
Portland Water District

North Carolina
Charlotte, City of, Stormwater Services

Pennsylvania
Philadelphia, City of

Tennessee
Chattanooga Stormwater Management

Texas
Harris County Flood Control District, Texas

Washington
Bellevue Utilities Department
Seattle Public Utilities

STATE

Arkansas Department of Environmental Quality
Connecticut Department of Environmental Protection
Kansas Department of Health & Environment
Kentucky Department of Environmental Protection
Ohio River Valley Sanitation Commission
Urban Drainage & Flood Control District, CO

CORPORATE

ADS Environmental Services
Advanced Data Mining International
Alan Plummer & Associates
Alden Research Laboratory Inc.
Alpine Technology Inc.
Aqua-Aerobic Systems Inc.
Aquateam–Norwegian Water Technology Centre A/S
ARCADIS
Associated Engineering
Black & Veatch
Boyle Engineering Corporation
Brown & Caldwell
Burgess & Niple, Ltd.
Burns & McDonnell
CABE Associates Inc.
The Cadmus Group
Camp Dresser & McKee Inc.
Carollo Engineers Inc.
Carpenter Environmental Associates Inc.
CDS Technologies Inc.
CET Engineering Services
Chemtrac Systems Inc.
CH2M HILL
Conestoga-Rovers & Associates
CONTECH Stormwater Solutions
D&B/Guarino Engineers, LLC
Damon S. Williams Associates, LLC
Earth Tech Inc.
Ecovation
EMA Inc.
Environ/The ADVENT Group, Inc.
Fay, Spofford, & Thorndike Inc.
Freese & Nichols Inc.
ftn Associates Inc.
Gannett Fleming Inc.
Garden & Associates, Ltd.
Geosyntec Consultants
GHD
Golder Associates Ltd.
Greeley and Hansen LLC
Hazen & Sawyer, P.C.
HDR Engineering Inc.
HNTB Corporation
Hydromantis Inc.
HydroQual Inc.
Infilco Degremont Inc.
Jacobson Helgoth Consultants Inc.
Jason Consultants LLC Inc.
Jordan, Jones, & Goulding Inc.
KCI Technologies Inc.
Kelly & Weaver, P.C.
Kennedy/Jenks Consultants
KMK Consultants
Komline Sanderson Engineering Corporation
Limno-Tech Inc.
Lombardo Associates Inc.
The Low Impact Development Center Inc.
Malcolm Pirnie Inc.
Material Matters
McKim & Creed
Metcalf & Eddy Inc.
Monteco Corporation
MPR Engineering Corporation, Inc.
MWH
NewFields Water Resources, LLC
O'Brien & Gere Engineers Inc.
Odor & Corrosion Technology Consultants Inc.
Oscar Larson & Associates
PA Government Services Inc.
Parametrix Inc.
Parsons
Post, Buckley, Schuh & Jernigan
Praxair, Inc.
RMC Water & Environment
Ross & Associates Ltd.
Rothberg, Tamburini & Windsor, Inc.
SAIC
Savin Engineers
Siemens Water Technologies
Stantec Consulting Inc.
Stearns & Wheler, LLC
Stone Environmental Inc.
Stratus Consulting Inc.
Synagro Technologies Inc.
Tetra Tech Inc.
Trojan Technologies Inc.
Trussell Technologies, Inc.
URS Corporation
Wade-Trim Inc.
Westin Engineering Inc.
Weston Solutions Inc.
Woodard & Curran
Zenon Environmental Inc.
Zoeller Pump Company

INDUSTRY

American Electric Power
American Water
ChevronTexaco Energy Research & Technology Company
The Coca-Cola Company
Dow Chemical Company
DuPont Company
Eastman Chemical Company
Eli Lilly & Company
Merck & Company Inc.
Premier Chemicals LLC
Procter & Gamble Company
Thames Water Utilities
Severn Trent Services Inc.
Suez Environnment
United Water Services LLC

Note: List as of 2/1/07

WERF Board of Directors

Chair
Vernon D. Lucy
Infilco Degremont Inc.

Vice-Chair
Dennis M. Diemer, P.E.
East Bay Municipal Utility District

Secretary
William J. Bertera
Water Environment Federation

Treasurer
James M. Tarpy, J.D.
Metro Water Services

Mary E. Buzby, Ph.D.
Merck & Company Inc.

Mohamed F. Dahab, Ph.D.
University of Nebraska, Lincoln

Charles N. Haas, Ph.D.
Drexel University

Robert W. Hite, J.D.
Metro Wastewater Reclamation District

Jerry N. Johnson
District of Columbia Water & Sewer Authority

Alfonso R. Lopez, P.E.
New York City Department of Environmental Protection

Peter J. Ruffier
City of Eugene

Murli Tolaney, P.E., DEE
MWH

R. Rhodes Trussell, Ph.D.
Trussell Technologies Inc.

Alan H. Vicory, Jr., P.E., DEE
Ohio River Valley Water Sanitation Commission

Rebecca F. West
Spartanburg Water

Executive Director
Glenn Reinhardt

WERF Research Council

Chair
Peter J. Ruffier
City of Eugene, Oregon

Vice-Chair
Karen L. Pallansch
Alexandria Sanitation Authority

Christine F. Andersen, P.E.
City of Long Beach, California

Gail B. Boyd
Independent Consultant

William C. Boyle, Ph.D.
University of Wisconsin

William L. Cairns, Ph.D.
Trojan Technologies Inc.

Glen T. Daigger, Ph.D.
CH2M HILL

Robbin W. Finch
Boise City Public Works

Ephraim S. King
U.S. EPA

Mary A. Lappin, P.E.
Kansas City, Missouri Water Services Department

Keith J. Linn
Northeast Ohio Regional Sewer District

Brian G. Marengo, P.E.
CH2M HILL

Drew C. McAvoy, Ph.D.
The Procter & Gamble Company

Margaret H. Nellor, P.E.
Nellor Environmental Associates, Inc.

Steven M. Rogowski, P.E.
Metro Wastewater Reclamation District of Denver

Michael W. Sweeney, Ph.D.
EMA Inc.

George Tchobanoglous, Ph.D.
Tchobanoglous Consulting

Gary Toranzos, Ph.D.
University of Puerto Rico

Ben Urbonas, P.E.
Urban Drainage and Flood Control District

IWA Publishing's authorised EU representative for General Product Safety Regulations is Diane D'Arras, 15 rue Duret, 75116 Paris, France, e-mail: safety@iwap.co.uk.

Printed and bound by CPI Group (UK) Ltd, Croydon, CR0 4YY
12/05/2026
02108865-0012